AF356448

AI-Based Transportation Planning and Operation

AI-Based Transportation Planning and Operation

Editor

Keemin Sohn

MDPI • Basel • Beijing • Wuhan • Barcelona • Belgrade • Manchester • Tokyo • Cluj • Tianjin

Editor
Keemin Sohn
Chung-Ang University
Korea

Editorial Office
MDPI
St. Alban-Anlage 66
4052 Basel, Switzerland

This is a reprint of articles from the Special Issue published online in the open access journal *Electronics* (ISSN 2079-9292) (available at: https://www.mdpi.com/journal/electronics/special_issues/transportation_planning).

For citation purposes, cite each article independently as indicated on the article page online and as indicated below:

LastName, A.A.; LastName, B.B.; LastName, C.C. Article Title. *Journal Name* **Year**, *Volume Number*, Page Range.

ISBN 978-3-0365-0364-6 (Hbk)
ISBN 978-3-0365-0365-3 (PDF)

Contents

About the Editor

Keemin Sohn was educated at the Department of Civil Engineering, Seoul National University (SNU), where, in 2003, he earned his Ph.D. in civil engineering. He is currently a professor with the Department of Urban Engineering at the Chung-Ang University, Seoul, Korea. His research interest encompasses the applications of artificial intelligence to transportation planning and operation. He has published more than 40 peer-reviewed research articles on machine-learning-based data-driven solutions to transportation-related problems.

Preface to "AI-Based Transportation Planning and Operation"

Due to drastic urbanization, cities around the world are facing transportation-induced problems such as congestion, accidents, and air pollution. Transportation planning and operation provide opportunities to satisfy the demand for the movement of people and goods in a safe, economical, convenient, and sustainable manner. Previous studies of transportation planning and operation have depended upon econometrics or engineering-based modeling, which cannot fully incorporate the power of data-driven or artificial intelligence (AI)-based approaches. Recently, AI technologies, such as supervised/unsupervised learning, reinforcement learning, and Bayesian modeling, have boosted the capability of dealing with the complexity and high nonlinearity in the problems of transportation planning and operations. More specifically, AI-based technologies in decision-making, planning, modeling, estimation, and control have facilitated the process of transportation planning and operations. This Special Issue provides an academic platform to publish high-quality articles on the applications of innovative AI algorithms to transportation planning and operation. The topics of the articles encompass AI technologies for traffic surveillance, vehicle emission reduction, traffic safety, congestion management, traffic speed forecasting, and ride sharing strategy.

Keemin Sohn
Editor

Article

Measuring Traffic Volumes Using an Autoencoder with No Need to Tag Images with Labels

Seungbin Roh, Johyun Shin and Keemin Sohn *

Department of Urban Engineering, Chung-Ang University, 84 Heukseok-ro, Dongjak-gu, Seoul 156-756, Korea; sbr444@cau.ac.kr (S.R.); olfy1021@cau.ac.kr (J.S.)
* Correspondence: kmsohn@cau.ac.kr

Received: 31 March 2020; Accepted: 24 April 2020; Published: 25 April 2020

Abstract: Almost all vision technologies that are used to measure traffic volume use a two-step procedure that involves tracking and detecting. Object detection algorithms such as YOLO and Fast-RCNN have been successfully applied to detecting vehicles. The tracking of vehicles requires an additional algorithm that can trace the vehicles that appear in a previous video frame to their appearance in a subsequent frame. This two-step algorithm prevails in the field but requires substantial computation resources for training, testing, and evaluation. The present study devised a simpler algorithm based on an autoencoder that requires no labeled data for training. An autoencoder was trained on the pixel intensities of a virtual line placed on images in an unsupervised manner. The last hidden node of the former encoding portion of the autoencoder generates a scalar signal that can be used to judge whether a vehicle is passing. A cycle-consistent generative adversarial network (CycleGAN) was used to transform an original input photo of complex vehicle images and backgrounds into a simple illustration input image that enhances the performance of the autoencoder in judging the presence of a vehicle. The proposed model is much lighter and faster than a YOLO-based model, and accuracy of the proposed model is equivalent to, or better than, a YOLO-based model. In measuring traffic volumes, the proposed approach turned out to be robust in terms of both accuracy and efficiency.

Keywords: autoencoder; deep learning; traffic volume; vehicle counting; CycleGAN

1. Introduction

Detecting traffic volumes is the basis on which traffic management and operation is implemented. For example, traffic signal control is totally dependent on the traffic volumes of each lane group that shares a signal phase. Measuring traffic volumes in real time is inevitable for a signal controller to adaptively assign signal phases to each lane group. Traffic volumes also determine the service level of a highway segment with respect to congestion management. For the purpose of transportation planning, an analyst depends on historical traffic volumes collected on an area-wide scale over a long-term basis to decide how best to budget for transportation infrastructure. A conventional spot detector, however, has limitations in accuracy and maintainability when implementing the aforementioned tasks. Although loop detectors are pervasive in current traffic surveillance [1,2], they cannot be a future option due to frequent breakdowns that entail large maintenance costs.

Traffic management centers that depend on probe vehicles also cannot measure traffic volumes and focus on collecting information on traffic speeds or travel times. The present study proposed a robust approach to constantly measure traffic volumes on an area-wide scale. Recently, computer vision technologies have replaced the spot-detecting and probe-based schemes, because high-resolution cameras are now more affordable and deep learning algorithms are easily accessible to engineers for use in processing images.

Prior to the success of deep learning, many rule-based algorithms had been developed to recognize vehicles within an image. The rule-based algorithms can be broken down into three categories according to taxonomy [3]. The first category uses the temporal difference method to capture moving objects by utilizing the differences in two consecutive images [4,5]. The performance of this method is largely affected by unexpected noises such as illumination drift due to changes in weather and to variations in the moving speed of objects. The second category involves use of the optical flow method to recognize moving vehicles from video frames by tracking changes in pixel intensities [6–9]. With this method, motion vectors of moving vehicles are mathematically derived from changes in pixel intensities. This method is also vulnerable when unexpected noise occurs and adds a computational burden to derive the motion vectors. The last category is background subtraction, which had been the most popular method before learning-based algorithms evolved [10,11]. This method utilizes the difference between a target image and the background image to recognize vehicles. Securing a consistent background image is a decisive factor in the success of this method. Some researchers adopted a dynamic background to avoid the impact of environmental changes by considering local features and motion histograms [12]. Even after obtaining a silhouette image from the background subtraction, a robust algorithm is necessary to recognize vehicles from the blob image. Two different approaches are available: Utilization of the convex hull theory [13] and devising an edge detector [14].

It should be noted that all these methods are rule-based rather than learning-based. This means that their models cannot evolve even when more image data are available. Recently, deep-learning algorithms such as YOLO v3 [15] and Fast-RCNN [16] scored a breakthrough in detecting objects within an image and have been successfully utilized in vehicle counting [17–19]. It is likely that learning-based algorithms will soon replace rule-based algorithms for traffic surveillance, since they incorporate advantages in both accuracy and maintainability. Such learning-based algorithms for vehicle detection, however, cannot accomplish the task of measuring traffic volumes. Traffic volume is defined as the number of vehicles passing a cross section of a road segment during a specified period such as 15 min or 1 h. It is necessary to track vehicles across time by continuously matching a vehicle detected in the previous video frame with that in the next frame. A standard Kalman filter, which assumes a constant velocity motion for the state equation and a linear relationship for the observation equation, has been adapted to track vehicles detected by YOLO [20]. The state vector is composed of the center coordinates, the height, the aspect ratio of a bounding box, and their velocities. The first three elements of the state vector are derived from direct observations of the vehicle state. The Kalman filter uses observations to iteratively predict and update the state vector.

Several algorithms are used to match the vehicle state forecasted by a Kalman filter with a corresponding observation in the next time frame. This matching problem can be solved via the Hungarian algorithm that uses the intersection over union (IOU) index [21]. Although the two-step approach succeeded in yielding a robust measurement performance, a simpler and lighter model must be developed.

Although a YOLO can be a robust vehicle detector, it requires a large computing resource for a field implementation. Furthermore, YOLO requires additional training whenever the testbed changes. Drawing a bounding box for every vehicle in training images entails considerable human effort, but it is required in order to accurately recognize vehicles in photos taken from different viewpoints [22].

In the present study, we devised an autoencoder model that would overcome the drawbacks of learning-based traffic volume counters. The proposed model requires neither human input to annotate images nor an additional algorithm to track vehicles after detection. The proposed model was trained in an unsupervised manner and thus eliminates the human effort needed to tag images with labels. In addition, the model is much lighter than any other learning-based models that are used to detect and track vehicles, because it utilizes only a small number of pixels within an image, which corresponds to a virtual cross line drawn on a road segment. The pixel intensities on the virtual line constitute a one-dimensional input vector, and the vector feeds an autoencoder as input and simultaneously becomes a label for the corresponding output. An autoencoder consists of two functional parts. While the former encoding part abstracts an input vector into features of a smaller dimension, the latter decoding part reproduces the original input vector from the abstracted features. Both parts have a generally symmetrical structure. The encoding part of the present model reduces the input vector into a scalar signal. After training, an autoencoder evaluated the signal from an input vector, and a simple rule was adopted to judge the presence of a vehicle. The latter decoding part is used only in the training stage to learn the parameters of an autoencoder.

As with other image-based object detection models, however, this autoencoder model was not immune to the impact of shadows and various illuminations. Furthermore, real vehicle images assume complex patterns, and thus signals representing the same vehicle are not consistent according to which part of a vehicle passes the detection line. Due to these complications, the signals were imperfect in judging whether a vehicle was occupying the virtual line sensor. To tackle the problem, original images were simplified using a CycleGAN developed by [23], so that both vehicles and backgrounds would have monotone colors. Such a transformation had already been successfully adopted in our previous studies measuring traffic speed and delay [22,24].

Section 2 explains how an autoencoder is used to recognize a vehicle's presence and shows how the model architecture is set up. Section 3 describes how to choose a testbed and collect image data. Section 4 shows how threshold values are selected to judge the vehicle presence and to count the vehicle passages. In the last subsection of Section 4, the traffic volumes measured from the present approach are compared with those from a YOLO+SORT algorithm [18]. Section 5 extends both the proposed model and the reference model, and the results are compared. The first subsection of Section 5 provides a brief description of the CycleGAN that was used to convert real photos into simple images. The following subsections describe how an autoencoder model was trained and tested on the synthesized images, and how a YOLO+SORT algorithm was fine-tuned using additional aerial photos. In Section 5, the efficiency of the proposed model is compared with a reference model and a plausible method to classify vehicle types based on the proposed model introduced. Finally, Section 6 draws conclusions and suggests further studies to develop the present approach.

2. An Autoencoder Recognizes the Presence of Vehicles

Autoencoders have been utilized to reduce the dimensions of input features [25,26], to remove the noise from corrupted input data [27,28], and to pretrain the weights of a supervised deep learning model [29,30]. In addition to these typical uses, autoencoders have been used for many other purposes [31]. An autoencoder belongs to the category of unsupervised learning models because it requires no human effort to label or annotate data for training. This is a great advantage of an autoencoder-based model when adapted to measuring traffic volumes. A standard autoencoder was employed in the present study to reduce the dimensions of an input feature into only a scalar signal without supervision. We expected the signal to be fully qualified to recognize the presence of a vehicle.

A cross line drawn on a road segment was assumed to be a virtual detector for counting traffic volumes. The width of the line was set as only a pixel, so that the input and output vectors for an autoencoder could be one-dimensional. Pixel intensities along the line vary according to the presence and absence of vehicles. The proposed approach was devised under the expectation that an autoencoder could recognize such differences in pixel intensities according to the presence or absence of a vehicle. The former encoder portion of the present autoencoder was established to be symmetrical with the latter decoder portion. Both the encoding and decoding portions of the autoencoder are composed of eight hidden layers, respectively. A single hidden node placed in the middle acts as a potential output node that produces a scalar signal to recognize the presence of a vehicle. The latter decoder portion is used only in the training stage.

$$Loss\ function = E\left[(F(X) - X)^{T} \cdot (F(X) - X)\right] \tag{1}$$

The discrepancy between the input and the output was set as the loss function to be minimized when training the autoencoder. The mean square error (MSE) was chosen to be the loss function. In Equation (1), X is a one-dimensional input vector of the autoencoder. In practice, the absolute value of pixel intensities subtracted from those of an empty road are used as input after normalization. $F(X)$ is the estimated output vector of the autoencoder. A stochastic gradient descent algorithm was used to train the model, and the batch size was 48. Training data were extracted from video frames of 100 min. Ten percent of the training data were randomly chosen to secure the convergence while sidestepping over-fitting.

Once the model training is complete, only the encoder portion computes a signal that can be used to recognize the presence of a vehicle. Dense layers are stacked to build both the encoder and decoder portions. The number of hidden nodes dwindles to 1 through the encoder portion and is then amplified up to the original input dimensions through the decoder portion. Figure 1 shows the entire architecture of the autoencoder adopted in the present study. For brevity, the activation using the rectified linear unit (ReLU) after each hidden layer is omitted in the figure. A scalar node in the middle is activated with a sigmoid function. The model architecture was set up after testing as many plausible alternative structures as possible.

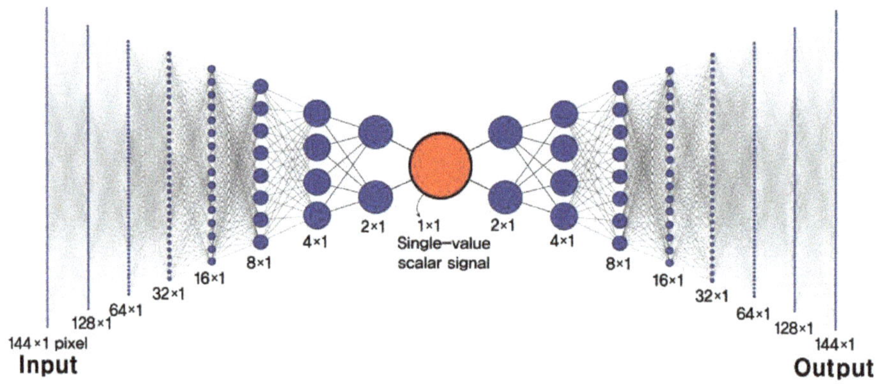

Figure 1. Architecture of the proposed autoencoder used to recognize the presence of a vehicle.

To facilitate the recognition of vehicle presence, the last signal of the encoder is activated by a sigmoid function, such that the signal ranges between 0 and 1. It is likely that the signal for vehicle presence will approach either of the boundary values. However, which value indicates the presence of a vehicle is unknown even after training the autoencoder. To understand the boundary value for vehicle presence, an analyst must intuitively examine the video stream used for training by comparing signal estimates with the vehicle presence. Based on the examination, a threshold value should be manually chosen to determine the presence or absence of vehicles. This engineering judgment could be the only limitation in the proposed approach. Although the threshold value may also vary according to the times of day and the weather conditions, it is difficult in practice to prepare different threshold values in advance. We suggest a revolutionary method in the fourth section to overcome this difficulty by transforming complex real photos into simple illustrations based on a CycleGAN.

3. Testbed and Data Collection

An intersection approach with four lanes was chosen as the testbed to train and test the proposed autoencoder. One hundred minutes of video stream were taken from 35 m above the intersection. Figure 2 shows a snapshot of the video stream. For each critical intersection in the Seoul metropolitan area, a 25–35 m high pole is planted to hang a CCTV for multipurpose surveillance. We assumed that the CCTV could be used for the purpose of traffic surveillance since it provides street photos from a bird's-eye view. The coverage of the testbed was 13 m wide and 34 m long. The virtual line detector was placed ahead of the stop line to avoid the condition whereby a vehicle would occupy the line during a red signal phase. Another reason to draw a virtual detection line closely to the stop line was because a lane change is not permitted when a vehicle is passing the stop line. Even though our ultimate goal was to measure traffic volumes, an autoencoder simply recognizes the presence or absence of a vehicle on a virtual detection line. How the autoencoder output is used to measure traffic volumes will be described in the next section.

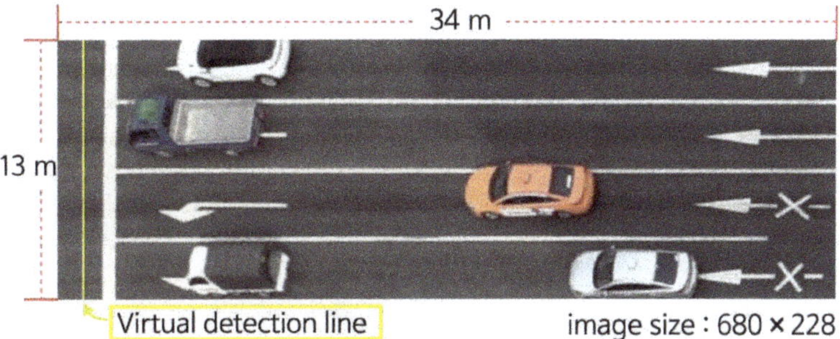

Figure 2. Testbed for the present study.

The proposed autoencoder model was trained and tested in a single testbed. Usually, testing results are most accurate when the model was trained in the same environment. It is rational that the proposed autoencoder should be trained for each site, since training the model is not a burdensome task without supervision. Trying to find a universal model that can cover all different locations is unnecessary for the present model. Basically, reducing the model size is more important for a real-time application than finding a heavier universal model with many parameters.

To assess the performance of the autoencoder, vehicle presence was manually identified during the entire time period. That is, the actual presence of vehicles on the virtual detection line was recorded every 0.2 s. Based on the ground truth, the precision and recall for the proposed autoencoder was computed (see the next section).

The present study was focused on confirming whether an autoencoder could be used to measure traffic volumes in the field on a real-time basis. To test the proposed approach, true traffic volumes were observed in the testbed. The number of vehicles passing the virtual detection line of each lane was counted for every two cycles of the traffic signal (=2.5 × 2 min). Based on these data, the final performance of the proposed approach was tested and compared with that of a reference model.

There are two types of object detectors based on deep learning technologies. A Fast-RCNN is accurate but has a disadvantage in computation time because it adopts a two-stage detection algorithm wherein a preprocess of regional proposals precedes a detection task. If a tracking stage is added when measuring traffic volumes, the model should go through three stages, which is not appropriate for a real-time application. On the other hand, a YOLO is a single-stage detection algorithm that has the merit of computation time, which has made the model widely accepted for the real-time recognition of objects. Since traffic volumes should be measured on a real-time basis, instead of the other two-stage models, we adopted a YOLO model as a reference to be compared with the proposed model. As mentioned earlier, the latest YOLO (version 3) with default weights is incapable of detecting vehicles in photos taken from a bird's-eye view. To fine-tune the YOLO model, we manually drew bounding boxes for all the vehicles in 1500 video frames of the testbed.

4. The Model Performance Based on Real Photos

All analyses in this section are based on real photos from video shoots. First, this section introduces a simple rule-based methodology to choose threshold values used to recognize the presence and absence of vehicles from the signal output of a trained autoencoder. Next, the precision and recall are computed for the autoencoder based on the chosen thresholds. Last, the performance of the proposed approach in measuring traffic volumes is compared with that of a "state-of-the-art" model based on YOLO and SORT [18].

4.1. Choosing Threshold Values to Measure Traffic Volumes

The trained autoencoder provided signals for the test data and the signal profile, as shown in Figure 3. Once a signal profile that a trained autoencoder emits is available, it is possible to find a threshold value with the naked eye by watching the video simultaneously with the signal profile. We selected a threshold value to recognize the presence of vehicles in such a way. When compared with actual video, signal values closer to 1 indicated that a vehicle was occupying the virtual detection line, whereas those closer to 0 corresponded to the background. Unfortunately, there is no rigorous way to drive an optimal boundary between the presence and absence of vehicles. We chose a threshold directly from the signal profile, such that it could be slightly larger than the lowest signal value for the background. The signal value for the background was almost 0, and the chosen threshold was set at 0.06. This manual choice did no harm in recognizing the presence of vehicles, which is verified in the next subsection by computing the precision and recall based on the ground truth.

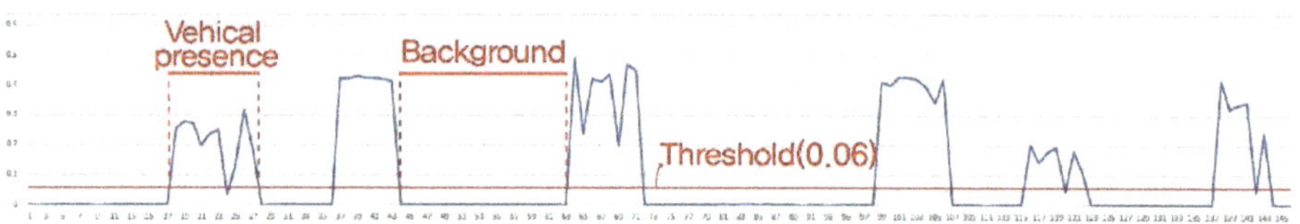

Figure 3. Signal profile from the autoencoder for the test data.

Although signals are derived from a sigmoid function, 0.5 cannot be selected as a threshold value because the signals corresponding to the presence of vehicles fluctuates significantly. The signal profile in Figure 3 shows variations in the signals of a vehicle's presence. This fluctuation could have been due to inconsistent shapes and colors of vehicles. If vehicle shapes and colors were consistent, it would be possible to establish a more robust way to choose a threshold value. A CycleGAN [23] will be adopted in the next section to transform real photos into simple illustrations with consistent vehicle shapes and colors.

Once a vehicle's presence is recognized, the raw signal profile can be updated in a clearer manner, as shown in Figure 4. Another threshold is the necessity to judge whether a series of signals belongs to a vehicle's passage. It is simple to count vehicles based on the updated signal profile, because the time gap between consecutive vehicles cannot be decreased to a minimum value in an intersection approach. Thus, a period of background shorter than the minimum gap was ignored and incorporated with the previous and next intervals of a vehicle's presence. The highway capacity manual (HCM, 2020) defines the saturation flowrate in an intersection approach for various geometric and operational conditions. The minimum headway derived from the saturation flowrate under ideal conditions is 1.8 s. This means that no vehicle could pass the stop line within 1.8 s right after the preceding vehicle has passed it. The minimum gap is shorter than the minimum headway according to the time it takes for the vehicle length to pass. The minimum gap (=0.5 s) chosen in the present study was very conservative. This minimum gap is globally applicable. However, the threshold value to recognize the vehicle presence is expected to vary from site to site.

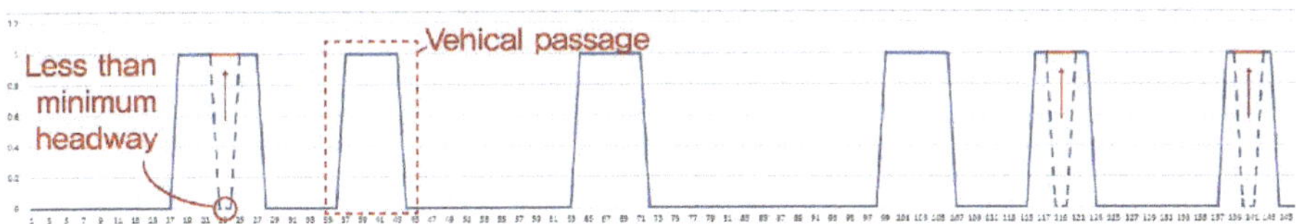

Figure 4. Final signal profile after judging the presence of vehicles based on chosen thresholds.

4.2. The Performance of the Autoencoder at Judging the Presence of Vehicles

The precision and recall of the trained autoencoder was computed to evaluate its performance for detecting the presence of vehicles. As mentioned earlier, to validate the autoencoder model we manually recognized the presence of vehicles. To facilitate the labeling task, only the video frames for green signals were utilized. Out of 100-min video frames, 80 min were used for training, and the remainder was reserved for testing. Table 1 shows the precision and recall for both the training and the testing data. Unlike the prior expectation, the testing results were ameliorated somewhat when compared with the training results. This might be due to a coincidence wherein the test dataset includes fewer problematic situations. The test results showed that the proposed autoencoder is not error free. A remedy was needed to increase the model performance, and that will be introduced in the next section.

Table 1. The precision and recall of the proposed autoencoder to identify the presence and absence of vehicles on the virtual detection line.

	Train Data			Test Data		
	Positive (Ground Truth)	Negative (Ground Truth)	Sum	Positive (Ground Truth)	Negative (Ground Truth)	Sum
Positive (Predicted)	19,806	456	20,262	4230	84	4314
Negative (Predicted)	1355	51,603	52,958	202	11,164	11,366
Sum	21,161	52,059	73,220	4432	11,248	15,680
Precision		0.977			0.981	
Recall		0.936			0.954	

4.3. The Performance of the Proposed Approach to Measure Traffic Volumes Compared with that of a Yolo-Based Model

The accuracy of measuring traffic volumes for the test data is introduced in this subsection. Since the autoencoder was not perfect in judging the presence of vehicles, its accuracy in measuring the traffic volumes entailed some errors. The traffic volume measured by the autoencoder-based methodology was overestimated when compared with observations. This implies that the methodology mistakenly double-counted some vehicle passages (Table 2).

A YOLO-based model was selected as a reference to validate the proposed model. The reference model adopted a two-step approach that involved detecting and tracking. The YOLO detects vehicles, and then a SORT algorithm tracks them [18]. Table 2 compares the accuracy of both the reference and the proposed methodology. Unexpectedly, the accuracy of the reference model was inferior to that of the proposed model. This resulted from the fact that the original YOLO had been trained only on vehicle images taken from the ground view, whereas testing images were taken from an aerial view. In the next section, the model will be fine-tuned using extra labeled images taken from an aerial view to enhance the performance. A CycleGAN-based remedy will also be used to allow the proposed model to overcome the inaccuracy.

Table 2. Comparing the accuracy of the proposed model with that of the reference model.

Accuracy Comparison		Autoencoder Model				YOLO v3 Model			
		Lane 1	Lane 2	Lane 3	Lane 4	Lane 1	Lane 2	Lane 3	Lane 4
C1~2	Predicted/ Ground truth	35/34	31/30	18/18	21/21	22/34	29/30	18/18	20/21
C3~4	Predicted/ Ground truth	39/39	42/40	17/17	19/19	38/39	39/40	14/17	17/19
C5~6	Predicted/ Ground truth	46/44	39/39	22/20	16/16	36/44	38/39	20/20	10/16
C7~8	Predicted/ Ground truth	41/43	45/46	21/20	18/18	31/43	44/46	32/20	13/18
Total	Predicted / Ground truth	161/160	157/155	78/75	74/74	127/160	150/155	84/75	60/74
	Error (%)	+0.6%	+1.3%	+4%	0%	−20.6%	−3.2%	+12%	−18.9%

5. The Model Performance Based on Synthesized Images

The dependence on real photos is an aspect of the proposed autoencoder that prevents it from sufficiently recognizing the presence of vehicles. Pixel intensities on the detection line are inconsistent while a vehicle is passing the line, because the shape and color of the vehicle varies. A robust way to change an original photo into a simpler image is introduced in this section under the assumption that the performance of the autoencoder could be enhanced if the shape and color of a vehicle were consistent within an image.

A CycleGAN was used to preprocess input images so that a vehicle image could be simpler and more consistent in both color and shape. It should be noted that a CycleGAN is not a prerequisite for an autoencoder to measure traffic volumes, but instead is an assistant tool to refine input images for the best accuracy.

5.1. The Introduction of a Cycle-Consistent Adversarial Network (CycleGAN)

The use of deep learning technologies in computer vision has recently trended toward the development of generative models to synthesize images [23,32,33]. In the use of these technologies, the approach of Zhu et al. [23] has been noteworthy in that an image in one context can be transformed to its corresponding image in another context. A CycleGAN has been used to change the photos of a summer landscape into those of a winter landscape, female photos into male photos, or satellite images into real physical maps. This capability of a CycleGAN also facilitates traffic surveillance that depends on images. We have already used images synthesized by a CycleGAN to enhance the accuracy of measuring both traffic speeds and delays [22,24]. This scheme was adopted in the present study to obtain images that were more consistent at recognizing the presence of vehicles on the detection line.

A CycleGAN is composed of two types of CNN models that are alternately updated during training. A generator converts images in one context to corresponding images in the other context, whereas a discriminator judges which context an image belongs to. Generators are trained to minimize the discrepancies between the images in one context and the images that are cycled from them. The training of discriminators must address two different objectives. A discriminator judges real photos as true and synthesized images as fake. Adversarial learning, however, attempts to deceive a discriminator so that it cannot discern a cycled image from the original real photo. For details concerning the training of a CycleGAN, readers can refer to previously published studies listed in the references [23,24].

Two separate image sets in different contexts are necessary to train a CycleGAN. In the present study, original photos were taken in the field, and cartoon-like illustrations were obtained from a traffic simulator (VISSIM v5.0, PTV, Karlsruhe, Germany). When training a CycleGAN two image sets do not have to be paired, which is a tremendous advantage. That means that no human effort is required to tag photos with labels in order to train a CycleGAN. The only requirement was traffic simulation that would mimic real traffic conditions. Most traffic simulators provide an animation module to visualize the simulation results. This function facilitates the generation of images that contain vehicles that are consistent in shape and color. Figure 5 shows examples of real photos and their corresponding images that were synthesized using a CycleGAN. The synthesized images include jelly-like vehicle shapes, each of which is represented by a constant color. In the next subsection, the enhancement in the performance of the autoencoder will be examined based on precision and recall.

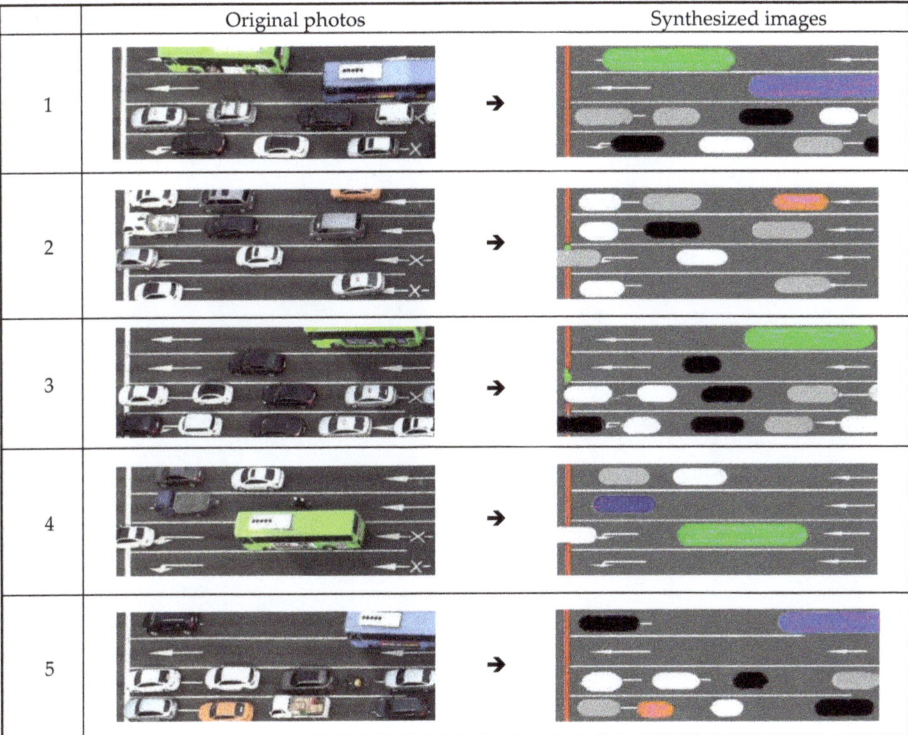

Figure 5. Original photos vs. synthesized images.

5.2. Enhancing the Performance of the Autoencoder to Better Judge the Presence of Vehicles

As expected, the performance of the autoencoder to judge the presence of vehicles was considerably improved when synthesized images were used instead of real photos for training and testing (compare Tables 1 and 3). The profile of raw signals from the enhanced autoencoder is quite different from that of the original autoencoder, which was based on real photos (see Figures 3 and 6). The updated profile more closely approximated the final profile after recognizing vehicle passages based on chosen thresholds (see Figures 6 and 7).

Table 3. The precision and recall of the autoencoder when synthesized images are used as input.

	Train Data			Test Data		
	Positive (Ground Truth)	Negative (Ground Truth)	Sum	Positive (Ground Truth)	Negative (Ground Truth)	Sum
Positive (Predicted)	20,051	376	20,427	4206	32	4238
Negative (Predicted)	942	51,851	52,793	213	11,229	11,442
Sum	20,993	52,227	73,220	4419	11,261	15,680
Precision		0.982			0.992	
Recall		0.955			0.952	

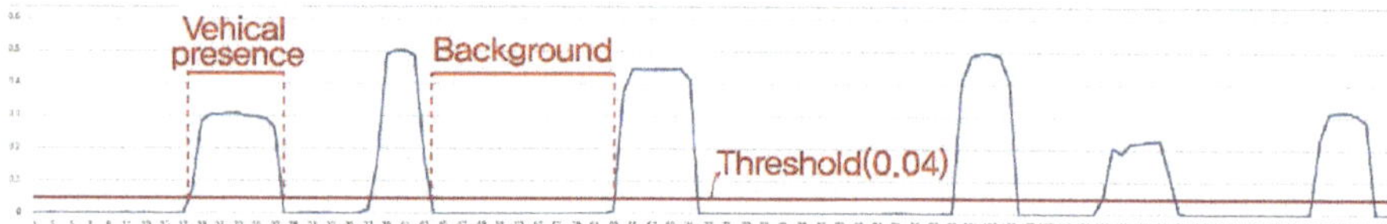

Figure 6. Updated signal profile from the autoencoder when using synthesized images for the test data.

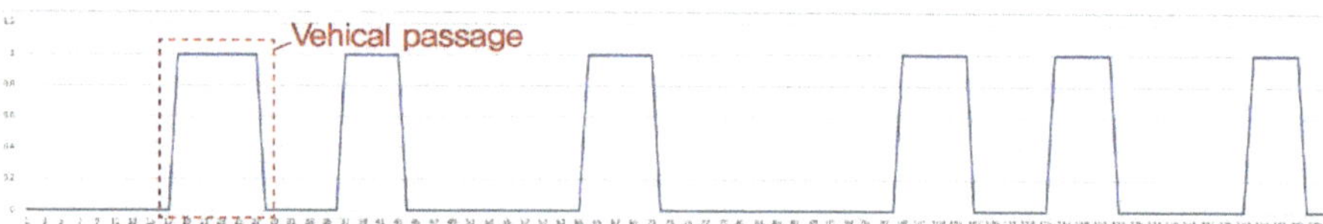

Figure 7. Final signal profile of vehicle passages when using synthesized images for the test data.

Figure 8 was drawn to directly show the reason for the performance enhancement. The signal of a real vehicle varies while passing the virtual detection line (=yellow line), whereas that of a jelly-type vehicle shape synthesized from a CycleGAN is much more consistent. This confirms that the proposed methodology has a great advantage when directly applied to field operation.

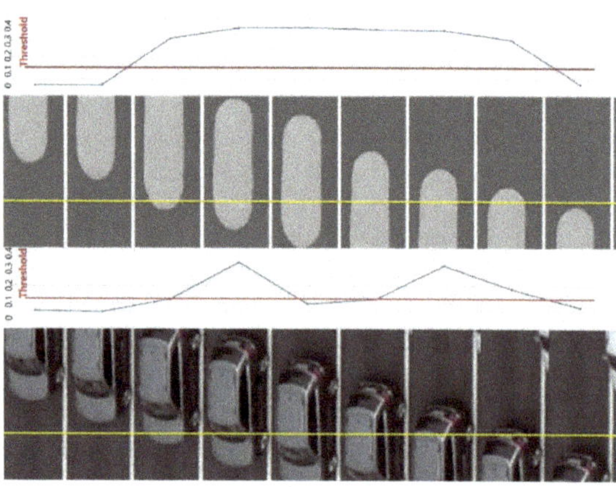

Figure 8. Signal profile comparison between a real vehicle (bottom) and a synthesized vehicle shape (top).

5.3. The Performance of the Updated Autoencoder in Measuring Traffic Volumes Compared with that of a Yolo-Based Model

The performance of the autoencoder to measure traffic volumes was improved considerably when synthesized images were used. The accuracy of the updated autoencoder approached 100%, which is much higher than that of the naive autoencoder trained on real photos (compare Tables 2 and 4).

Table 4. The accuracy comparison between the updated model and reference model.

Accuracy Comparison		Autoencoder Model (Updated)				YOLO v3 Model (Fine Tuned)			
		Lane 1	Lane 2	Lane 3	Lane 4	Lane 1	Lane 2	Lane 3	Lane 4
C1~2	Predicted/Ground truth	34/34	30/30	18/18	21/21	34/34	20/20	18/18	21/21
C3~4	Predicted/Ground truth	39/39	40/40	17/17	19/19	47/39	40/40	16/17	19/19
C5~6	Predicted/Ground truth	44/44	39/39	20/20	16/16	44/44	39/39	20/20	16/16
C7~8	Predicted/Ground truth	41/43	45/46	20/20	18/18	41/43	45/46	25/20	16/18
Total	Predicted/Ground truth	158/160	154/155	75/75	74/74	166/160	154/155	79/75	72/74
	Error (%)	−1.3%	−0.6%	0%	0%	+3.8%	−0.6%	+5.3%	−2.7%

The performance of the updated autoencoder to measure traffic volumes was compared with that of a YOLO-based model. For a fair comparison, a YOLO-based model was also updated with extra labeling data. One thousand and five hundred pictures from an aerial view were annotated with bounding boxes to fine tune the original YOLO v3 model with default weights. This took almost a week with two coding experts hired. Drawing a bounding box for every vehicle within all images is a very labor-intensive task, which makes the use of a YOLO-based model less practical for use in traffic surveillance. The performance of the YOLO-based model after fine tuning was comparable to that of the proposed model (Table 4). It should be noted that the proposed model does not require human effort for the task of labeling. Both the autoencoder and the CycleGAN belong to the category of unsupervised learning.

6. The Model Efficiency and the Potential to Classify Vehicle Types

One of the main purposes of the present study is to develop a lighter model that could be implemented on a real-time basis using edge computing. A one-dimensional input instead of a full image has a great advantage in saving computer memory onsite. The comparison of computing time and memory usage is introduced in Table 5. The number of parameters in the proposed model was much smaller than that in a YOLO. The proposed model turned out to be much lighter and faster than a YOLO when recognizing vehicles for a frame. Even when a CycleGAN was used to preprocess an input image, a YOLO did not outperform the proposed model in computing speed.

Table 5. The efficiency comparison between the proposed model and reference model.

	Number of Parameters	Evaluation Time per Frame
Autoencoder model	59,361	0.00007 s
CycleGAN + Autoencoder model	9,525,101	0.032 s
YOLO v3 model	61,581,727	0.078 s

The proposed model did not include the intrinsic function of classifying vehicle types, whereas other image-based models such as YOLO can easily classify vehicle types. A plausible methodology is proposed to classify vehicle types via a simple rule-based approach. It is possible that the vehicle length could be approximately measured under the assumption that the length would be proportional to the passing time. This assumption has been widely accepted when the vehicle type was estimated based on signals from a conventional loop detector. The moving average of a vehicle's passage time was dynamically updated and used as a basis on which the vehicle length was determined. If a vehicle passage time is larger than a threshold value, the vehicle is categorized into a group of large trucks or buses (Figure 9). We determined a threshold value to be 1.67-fold of the average of vehicle passage times.

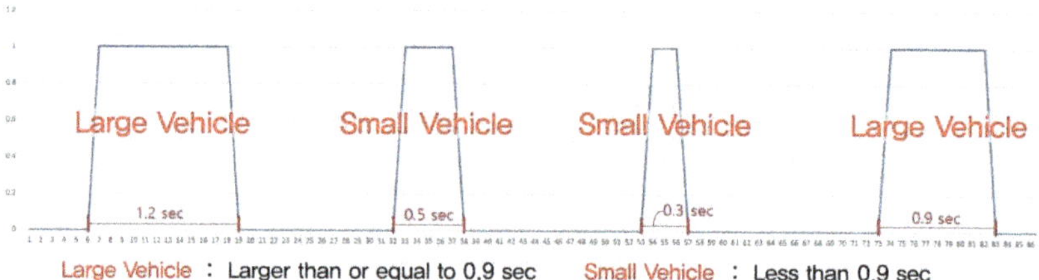

Figure 9. Threshold to classify vehicle types.

On the other hand, a YOLO has the ability to directly classify vehicle types. We reduced the number of classes of a YOLO model and trained it from scratch using our own labeled images without depending on pretrained parameters. A YOLO version 3 was trained so that it could classify three different types of vehicles such as cars, small vehicles, and large vehicles. Small vehicles include SUVs and small trucks, and large vehicles include buses, large trucks, and trailers. For the comparison with the proposed model with only two classes, cars and small vehicles were categorized into a single group of small vehicles. As shown in Table 6, the proposed model turned out to be capable of classifying vehicle types, even though the performance was slightly inferior to that of a YOLO.

The motorcycles were excluded in measuring traffic volumes. The passage of a motorcycle left smaller signals than vehicles. These signals of motorcycles could be a cause of false positives. Although a carefully chosen threshold turned out to minimize the errors, rigorous measures should be taken to recognize motorcycles, which will depend on two-dimensional signals in the further study.

Table 6. The accuracy comparison for classifying vehicle types.

Accuracy Comparison	Autoencoder				CycleGAN + Autoencoder				YOLO v3			
	Lane 1	Lane 2	Lane 3	Lane 4	Lane 1	Lane 2	Lane 3	Lane 4	Lane 1	Lane 2	Lane 3	Lane 4
Small Vehicle	152/150	155/155	74/73	71/72	149/150	154/155	73/73	72/72	156/150	154/155	77/73	70/72
Large Vehicle	9/10	2/0	4/2	3/2	9/10	0/0	2/2	2/2	10/10	0/0	2/2	2/2

7. Conclusions

A novel methodology to measure traffic volumes without supervision was developed in the present study. An autoencoder was trained to recognize the presence of vehicles on a cross section of a roadway based solely on pixel intensities. The proposed methodology requires no human effort to tag images with labels. The performance of the proposed model trained on synthesized images approximated that of a YOLO-based model that was fine-tuned with the extra labeling of images. When considering the human effort required for the labeling task, the proposed methodology seems more promising for use in the field.

The proposed methodology, however, demonstrated a critical drawback wherein vehicle types could not be distinguished. It is possible to approximately classify vehicle types if a constant speed is assumed for all moving vehicles. This assumption is acceptable when the detection line is placed on the stop line of an intersection approach. Speed estimation that depends on spot detectors usually adopts such an assumption. Further study is expected to develop a traffic volume counter that could classify each vehicle type.

Author Contributions: Conceptualization, K.S.; Data curation, J.S.; Formal analysis, S.R.; Writing—original draft, K.S. All authors have read and agreed to the published version of the manuscript.

Funding: This research was supported by the Chung-Ang University Research Scholarship Grants in 2019 and also by the Korea Agency for Infrastructure Technology Advancement (KAIA) grant funded by the Ministry of Land, Infrastructure, and Transport (grant number 19TLRP-B148659-02).

Conflicts of Interest: The authors declare no conflicts of interest.

References

1. Riedel, T.; Erni, K. Do my loop detectors count correctly? A set of functions for detector plausibility testing. *J. Traffic Transp. Eng. (JTTE)* **2016**, *4*, 117–130.
2. Meng, C.; Yi, X.; Su, L.; Gao, J.; Zheng, Y. City-wide traffic volume inference with loop detector data and taxi trajectories. In Proceedings of the 25th ACM SIGSPATIAL International Conference on Advances in Geographic Information Systems, Redondo Beach, CA, USA, 7–10 November 2017; pp. 1–10.
3. Ozkurt, C.; Camci, F. Automatic Traffic Density Estimation and Vehicle Classification for Traffic Surveillance System using Neural Networks. *Math. Comput. Appl.* **2009**, *14*, 187–196. [CrossRef]
4. López, M.T.; Fernández-Caballero, A.; Mira, J.; Delgado, A.E.; Fernández, M.A. Algorithmic lateral inhibition method in dynamic and selective visual attention task: Application to moving objects detection and labeling. *Expert Syst. Appl.* **2006**, *31*, 570–594. [CrossRef]
5. López, M.T.; Fernández-Caballero, A.; Fernández, M.A.; Mira, J.; Delgado, A.E. Visual Surveillance by Dynamic Visual Attention Method. *Pattern Recogn.* **2006**, *39*, 2194–2211. [CrossRef]
6. Ji, X.; Wei, Z.; Feng, Y. Effective vehicle detection technique for traffic surveillance systems. *J. Vis. Commun. Image Represent.* **2006**, *17*, 647–658. [CrossRef]
7. Qimin, X.; Xu, L.; Mingming, W.; Bin, L.; Xianghui, S. A methodology of vehicle speed estimation based on optical flow. In Proceedings of the 2014 IEEE International Conference on Service Operations and Logistics, and Informatics, Qingdao, China, 8–10 October 2014; pp. 33–37.
8. Indu, S.; Gupta, M.; Bhattacharyya, A. Vehicle tracking and speed estimation using optical flow method. *Int. J. Eng. Sci. Technol.* **2011**, *3*, 429–434.
9. Haag, M.; Nagel, H.H. Combination of edge element and optical flow estimates for 3D-model-based vehicle tracking in traffic image sequences. *Int. J. Comput. Vis.* **1999**, *35*, 295–319. [CrossRef]
10. Zhou, J.; Gao, D.; Zhang, D. Moving vehicle detection for automatic traffic monitoring. *IEEE Trans. Veh. Technol.* **2007**, *56*, 51–59. [CrossRef]
11. Niu, X. A semi-automatic framework for highway extraction and vehicle detection based on a geometric deformable model. *ISPRS J. Photogramm. Remote Sens.* **2006**, *61*, 170–186. [CrossRef]
12. Qian, Z.M.; Shi, H.X.; Yang, J.K. Video vehicle detection based on local feature. *Adv. Mater. Res.* **2011**, *186*, 56–60. [CrossRef]
13. Jayaram, M.A.; Fleyeh, H. Convex hulls in image processing: a scoping review. *Am. J. Intell.Syst.* **2016**, *6*, 48–58.
14. Zhu, Z.; Xu, G. VISATRAM: A real-time vision system for automatic traffic monitoring. *Image Vis. Comput.* **2000**, *18*, 781–794. [CrossRef]
15. Redmon, J.; Divvala, S.; Girshick, R.; Farhadi, A. You only look once: Unified, real-time object detection. In Proceedings of the IEEE conference on Computer Vision and Pattern Recognition(CVPR), Las Vegas, NV, USA, 6 June–1 July 2016; pp. 779–788.
16. Ren, S.; He, K.; Girshick, R.; Sun, J. Faster r-cnn: Towards real-time object detection with region proposal networks. In Proceedings of the Advances in Neural Information Processing Systems (NIPS), Montreal, QC, Canada, 7–12 December 2015; pp. 91–99.
17. Asha, C.S.; Narasimhadhan, A.V. Vehicle counting for traffic management system using YOLO and correlation filter. In Proceedings of the 2018 IEEE International Conference on Electronics, Computing and Communication Technologies (CONECCT), Bangalore, India, 16–17 March 2018; pp. 1–6.

18. Lopez, G. Traffic Counter with YOLO and SORT. Available online: https://github.com/guillelopez/python-traffic-counter-with-yolo-and-sort (accessed on 11 December 2018).
19. Sang, J.; Wu, Z.; Guo, P.; Hu, H.; Xiang, H.; Zhang, Q.; Cai, B. An improved YOLOv2 for vehicle detection. *Sensors* **2018**, *18*, 4272. [CrossRef] [PubMed]
20. Bathija, A.; Sharma, G. Visual Object Detection and Tracking using YOLO and SORT. *Int. J. Eng. Res. Technol. (IJERT)* **2019**, *8*, 705–708.
21. Wojke, N.; Bewley, A.; Paulus, D. Simple online and realtime tracking with a deep association metric. In Proceedings of the 2017 IEEE International Conference on Image Processing (ICIP), Beijing, China, 17–20 September 2017; pp. 3645–3649.
22. Shin, J.; Roh, S.; Sohn, K. Image-Based Learning to Measure the Stopped Delay in an Approach of a Signalized Intersection. *IEEE Access* **2019**, *7*, 169888–169898. [CrossRef]
23. Zhu, J.Y.; Park, T.; Isola, P.; Efros, A.A. Unpaired image-to-image translation using cycle-consistent adversarial networks. *arXiv* **2017**, arXiv:1703.10593.
24. Lee, J.; Roh, S.; Shin, J.; Sohn, K. Image-based learning to measure the space mean speed on a stretch of road without the need to tag images with labels. *Sensors* **2019**, *19*, 1227. [CrossRef] [PubMed]
25. Zabalza, J.; Ren, J.; Zheng, J.; Zhao, H.; Qing, C.; Yang, Z.; Du, P.; Marshall, S. Novel segmented stacked autoencoder for effective dimensionality reduction and feature extraction in hyperspectral imaging. *Neurocomputing* **2015**, *185*, 1–10. [CrossRef]
26. Hinton, G.E.; Salakhutdinov, R. Reducing the dimensionality of data with neural networks. *Science* **2016**, *313*, 504–507. [CrossRef] [PubMed]
27. Lu, X.; Tsao, Y.; Matsuda, S.; Hori, C. Speech enhancement based on deep denoising autoencoder. In Proceedings of the Interspeech, Lyon, France, 25–29 August 2013; pp. 436–440.
28. Vincent, P.; Larochelle, H.; Bengio, Y.; Manzagol, P.A. Extracting and composing robust features with denoising autoencoders. In Proceedings of the 25th international conference on Machine learning (ICML), Association for Computing Machinery, New York, NY, USA, 5–9 July 2008; pp. 1096–1103.
29. Pasa, L.; Sperduti, A. Pre-training of recurrent neural networks via linear autoencoders. In Proceedings of the Advances in Neural Information Processing Systems, Montreal, QC, Canada, 8–13 December 2014; pp. 3572–3580.
30. Glüge, S.; Böck, R.; Wendemuth, A. Auto-encoder pre-training of segmented-memory recurrent neural networks. In Proceedings of the European Symposium on Artificial Neural Networks (ESANN), Bruges, Belgium, 24–26 April 2013.
31. Goodfellow, I.; Bengio, Y.; Courville, A. *Deep Learning*; MIT Press: Cambridge, UK, 2016.
32. Goodfellow, I.; Pouget-Abadie, J.; Mirza, M.; Xu, B.; Warde-Farley, D.; Ozair, S.; Courville, A.; Bengio, Y. Generative adversarial nets. In Proceedings of the Advances in Neural Information Processing Systems, Montreal, QC, Canada, 8–13 December 2014; pp. 2672–2680.
33. Radford, A.; Metz, L.; Chintala, S. Unsupervised representation learning with deep convolutional generative adversarial networks. *arXiv* **2015**, arXiv:1511.06434.

Article

Bottleneck Based Gridlock Prediction in an Urban Road Network Using Long Short-Term Memory

Ei Ei Mon [1], Hideya Ochiai [2], Chaiyachet Saivichit [1] and Chaodit Aswakul [1,*]

[1] Wireless Network and Future Internet Research Unit, Department of Electrical Engineering, Faculty of Engineering, Chulalongkorn University, Bangkok 10330, Thailand; 6071456921@student.chula.ac.th (E.E.M.); chaiyachet.s@chula.ac.th (C.S.)

[2] Information and Communication Engineering, Graduate School of Information Science and Technology, The University of Tokyo, Tokyo 113-8656, Japan; jo2lxq@hongo.wide.ad.jp

* Correspondence: chaodit.a@chula.ac.th; Tel.: +66-087-693-6390

Received: 13 July 2020; Accepted: 25 August 2020; Published: 1 September 2020

Abstract: The traffic bottlenecks in urban road networks are more challenging to investigate and discover than in freeways or simple arterial networks. A bottleneck indicates the congestion evolution and queue formation, which consequently disturb travel delay and degrade the urban traffic environment and safety. For urban road networks, sensors are needed to cover a wide range of areas, especially for bottleneck and gridlock analysis, requiring high installation and maintenance costs. The emerging widespread availability of GPS vehicles significantly helps to overcome the geographic coverage and spacing limitations of traditional fixed-location detector data. Therefore, this study investigated GPS vehicles that have passed through the links in the simulated gridlock-looped intersection area. The sample size estimation is fundamental to any traffic engineering analysis. Therefore, this study tried a different number of sample sizes to analyze the severe congestion state of gridlock. Traffic condition prediction is one of the primary components of intelligent transportation systems. In this study, the Long Short-Term Memory (LSTM) neural network was applied to predict gridlock based on bottleneck states of intersections in the simulated urban road network. This study chose to work on the Chula-Sathorn SUMO Simulator (Chula-SSS) dataset. It was calibrated with the past actual traffic data collection by using the Simulation of Urban MObility (SUMO) software. The experiments show that LSTM provides satisfactory results for gridlock prediction with temporal dependencies. The reported prediction error is based on long-range time dependencies on the respective sample sizes using the calibrated Chula-SSS dataset. On the other hand, the low sampling rate of GPS trajectories gives high RMSE and MAE error, but with reduced computation time. Analyzing the percentage of simulated GPS data with different random seed numbers suggests the possibility of gridlock identification and reports satisfying prediction errors.

Keywords: bottleneck and gridlock identification; gridlock prediction; urban road network; long short-term memory

---

1. Introduction

Traffic congestion has been an increasing problem in the busiest urban cities. The main obstacles of traffic congestion are illegal parking, road maintenance, lane closure due to utility work, narrowing roads, accidents, and weather conditions. These incidences lead to traffic bottlenecks, which cause several adverse effects on the number of crashes, the increased cost of travelers and commuters, and fuel consumption. The traffic bottlenecks can often be located at strategic locations in a network, e.g., at off-ramps, on-ramps, and lane drop areas [1]. Practically, there have been various useful definitions of bottlenecks proposed by several authors identifying traffic fluctuations along with connected roadway segments. A bottleneck implies the congestion evolution and queue formation, which consequently disturb travel delay and worsen the urban traffic environment and safety [2]. In our previous work [3], a bottleneck was defined in that the demand and supply are mismatched due to the network structure such as upstream links merging to only one downstream link in each intersection area.

Traffic management and control systems have become increasingly important for developing traffic operations' efficiency and safety [4]. To improve traffic management, various sources of data are available via different traffic data collection methods. A series of spot sensors such as inductive loop detectors, remote transportation microwave sensors, and video cameras can provide accurate traffic information. However, these fixed-location sensors can only incorporate the traffic state at specific locations, and these devices might include high maintenance expenditure with frequent malfunction. For urban road networks, sensors need to cover a wide range of areas, especially for bottleneck and upstream/downstream analysis. The high installation and maintenance costs of those sensors are proven to be impractical for many cities.

Compared to the traditional way of using fixed-location sensors, with further advances in technology, using vehicles equipped with Global Positioning System (GPS) sensors as probes to collect traffic data has become popular. Ubiquitous traffic data are available everywhere automatically, and this helps develop Intelligent Transportation Systems (ITSs). The emerging widespread availability of probe data significantly helps to overcome the geographic coverage and spacing restrictions of traditional loop detector data [5]. Using such vehicles can provide real-time information on the traffic conditions along with the entire road network. These GPS-equipped vehicles can collect and output mobility data periodically, including longitude, latitude, speed, vehicle headings, and timestamps.

Traffic condition prediction is one of the primary components of ITSs, and its attention has grown in transportation research. The objective is to provide individual commuters or travelers with accurate traffic information on time. Some of the most promising predictions could improve travel time reliability with the accurate prediction for the same trips compared with the same day of adjacent weeks because time series of traffic data usually have recurrent temporal patterns. For example, similar traffic congestion happens every morning and evening rush hour. Further, this pattern is likely to occur weekly, monthly, and yearly.

This research aims at identifying bottleneck at each intersection and predicting urban network gridlocks. To summarize, the primary contributions of this paper are listed as follows:

1. This paper proposes a bottleneck definition in an urban traffic network according to the congestion state of each intersection;
2. This paper proposes a gridlock definition in the urban traffic network according to the bottleneck states at each intersection in the gridlock-looped intersections;
3. The difference from the previous work [3] is in trying to use GPS vehicles to gain advantages over the high installation and maintenance costs of traditional detectors in practice. Concerning the difficulty of obtaining the GPS mobility data due to privacy issues, in this research, the vehicles traveling around the point of interest area in the simulated urban network are assumed as GPS vehicles. However, the critical concern with the limited coverage by GPS vehicles is to study the required minimum number of GPS vehicles while maintaining the desired precision and accuracy levels [6]. Therefore, this work investigates 1% to 50% available GPS vehicles passing through the links in the simulated gridlock area. For the development of bottleneck and gridlock prediction, we chose the Chula-SSS (Chula-Sathorn SUMO Simulator) dataset [7], which has already been calibrated with morning and evening cases for the Sathorn road network area, for our study;
4. This paper proposes gridlock prediction using temporal time series analysis based LSTM with time-lagged observations. This paper considers the lowest 1 min time-lagged observation period as a baseline to compare the prediction error values to study the effectiveness of LSTM on extended time-lagged observations.

The remainder of the paper is organized as follows. Section 2 provides the related works. Section 3 presents the simulation framework. Section 4 proposes the methodology of traffic data processing, the description of bottleneck and gridlock identification, and the study of sample sizes for gridlock detection. Section 5 presents the urban gridlock prediction based on the bottleneck with LSTM. Section 6 discusses the experimental results. Finally, concluding remarks are presented in Section 7.

2. Related Works

Traffic control and management are a primary issue in urban transport networks. In general, bottlenecks are identified and studied to find ways to overcome their negative effects on traffic. The traffic bottlenecks in urban road networks are more challenging to investigate than in freeways or simple arterial networks [8]. Traffic bottlenecks in urban road networks vary with traffic demands with spatiotemporal characteristics. Other events, such as traffic signals, traffic accidents, roadworks, and severe weather conditions, are required to consider more of their impacts on urban networks than in freeways.

When a bottleneck happens, there is a notable difference between traffic speed in the bottleneck area and its nearby downstream links. Oversaturation is usually caused by vehicle queues first occurring at some bottleneck intersections (i.e., critical intersections) and then propagating to neighboring upstream and downstream links between intersections. In [9], to identify oversaturated intersections, the authors proposed residual queue length estimation using shockwave speed. The authors also presented the detection of spill-over by identifying long detector occupancy time during the green phase. In [10], the authors proposed the identification of bottlenecks in an urban area practicing on causal congestion trees and graphs with the correlations of road segments according to the congestion propagation speed. More recently, in [11], the authors studied a traffic bottleneck identification method based on the traffic speed of loop detector data using the fusion of different collection cycles of loop detector data. To identify urban bottlenecks, in [1], the authors proposed a congestion analysis on the bottleneck core link and its neighboring links by clustering links based on the ranks of speed.

More recently, GPS-equipped vehicles were used in the field of bottleneck identification. In [12], the authors presented the processes of identifying and defining bottlenecks using GPS data obtained by trucks. In [13], the authors explored and identified bottlenecks in urban areas based on recurrent low-speed segments in the road network using GPS data. In [14,15], the authors used a spatiotemporal variation of speed to identify and classify bottlenecks using GPS data and crowdsourced traffic data. In [16], the authors tried to identify recurrent bottlenecks using spatial and temporal concepts on probe-reported speed data. In [17], the authors proposed a bottleneck identification method on urban expressways based on the traffic speed variance between the bottleneck area and its neighboring downstream links by analyzing the floating car data. In [18], the authors investigated traffic state and identified traffic bottlenecks on urban expressways using the fusion data accessible from fixed detectors and mobile navigation application data.

Future traffic condition prediction is a general concept [19]. In [20], the authors proposed the characterization of the traffic predictor to analyze vehicular traffic behavior for the different road segments using SUMO. In recent years, as deep learning techniques have advanced rapidly, researchers have started to adopt deep neural networks for high-accuracy traffic prediction. The Recurrent Neural Network (RNN) is widely known as a proper method to recognize the spatiotemporal evolution of traffic flow. In [21], the authors proposed a deep learning architecture based on the Recurrent Neural Network and Restricted Boltzmann Machine (RNN-RBM) to predict the spatiotemporal congestion evolution pattern using GPS data. In [22], the authors used an error-feedback Recurrent Convolutional Neural Network structure (eRCNN) for continuous traffic speed prediction. However, the traditional RNN fails to capture the long-term evolution. To resolve the vanishing gradient problem in RNN, a Long Short-Term Memory (LSTM) network was proposed in [23]. LSTM is well suited to classify, process, and predict time series given time lags of unknown duration. In [24], the authors proposed a novel short-term traffic volume prediction model using

LSTM. More recently, in [25], the authors proposed a multiple time step short-term traffic prediction architecture using the RNN framework based on LSTM.

As mentioned above, the studies discussed identifying bottlenecks in intersections and respective upstream and downstream links in urban areas. Furthermore, LSTM has been applied in short-term traffic prediction. However, to the best of our knowledge, research on gridlock analysis has not yet been fully studied. We hope that this research will fill this gap by discussing the bottleneck and gridlock analysis based on the bottleneck at each intersection in the urban area. Congestion from the bottleneck originates traffic jams and propagates to neighboring upstream and downstream vicinities [26]. Such a situation usually leads to gridlock, which can potentially reduce traffic efficiency to an almost halted state, especially in a complex urban road network. Gridlock is used to describe severe road traffic congestion with zero flow [27,28]. Each interrelated intersection in a road network is fully occupied by slow-speed vehicles, and vehicles on conflicting approaches cannot move forward even when receiving a green traffic light, as shown in Figure 1. In our previous work [3], we proposed gridlock detection based on recurrent and non-recurrent congestion using the measurable gridlock characteristics in terms of traffic jam length and speed for both the upstream and downstream links of corresponding intersections. In previous work, we used simulated lane area detectors to detect gridlock, but these fixed-location detectors have some spacing limitations and installation costs in practice, as mentioned above.

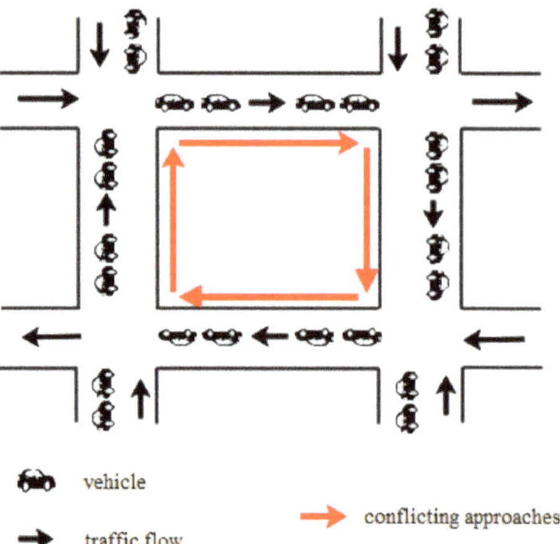

Figure 1. An example of gridlock-looped intersections.

3. Simulation Framework

Simulation models that can closely represent the real-world scenario are powerful and useful. In the area of road traffic simulation, three different models [29] are used, i.e., the (1) microscopic model, (2) macroscopic model, and (3) mesoscopic model. Microscopic simulation models have been used in several transportation areas in developing ITS technologies.

In this research, the Simulation of Urban MObility (SUMO) has been used to handle large, complicated road networks at a microscopic (vehicle-level) scale. SUMO is an open-source microscopic simulator developed by the German Aerospace Centre DLR in 2001 [30]. SUMO supports the traffic simulation community with a full-featured suite of modeling utilities, including the TraCI tool [31]. This tool is a Python API allowing users during the simulation run-time to retrieve simulated objects' values, e.g., in accessing the speed of simulated GPS vehicles on each link in the simulated urban network.

For traffic police and traffic engineer deployment, an educational tool [7] has been introduced by the Chulalongkorn University's Sathorn Model project in Bangkok (Thailand), namely the Chula-Sathorn SUMO Simulator (Chula-SSS). By using this tool, traffic police can see the traffic signal phase, the queue length of vehicles in each upcoming direction at each signalized intersection, and the overall area with neighborhood road segments with user-friendly interfaces. Chula-SSS supports two calibrated datasets (morning and evening rush hours) for the Sathorn road network area. This study chose to investigate the morning case of Chula-SSS from 6 am to 9 am (corresponding to 21,600 s to 32,400 s after midnight). This time interval is the morning congestion period during weekdays [7], whereby in practice, the local traffic police have often observed the real occurrences of gridlocks in the network. The simulated Chula-SSS consists of 2375 intersection nodes, 4517 edges, and 10 signalized intersections, as shown in Figure 2.

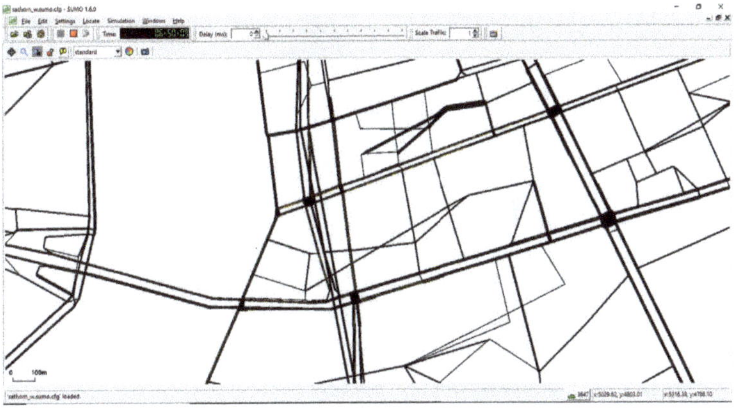

Figure 2. Sathorn road network topology in SUMO.

The dataset consists of 55,000 traveling vehicles in the morning or evening rush hours. Recurrent based gridlock happens on a daily basis in both morning and evening rush hours. This gridlock phenomenon has been intentionally disabled during the calibration attempt of Chula-SSS to find the optimal mobility model parameters. To exhibit the case of gridlock during morning rush hours, the previous work [3] already included two extra routes (1 and 2) to the Sathorn critical region of Chula-SSS with traffic flows of 2125 vehicles (Route 1) and 1500 vehicles (Route 2) per hour, with their routes as shown in Figure 3. The Sathorn critical region is where about 12,000 out of 55,000 vehicles travel on 10 edges, as shown in Figure 4. The GPS data collection has different time resolutions (1 s, 5 s, 10 s, 15 s, 20 s, 25 s, 30 s, 35 s, 40 s, 45 s, 50 s, 55 s, 60 s) within the simulation time interval for three hours from 6 am to 9 am. The collected vehicle data are assumed as GPS vehicles. With the limited penetration ratio of GPS vehicles, the different number of samples of GPS vehicles (1%, 5%, 10%, 15%, 20%, 25%, 30%, 35%, 40%, 45%, 50%) that passed through the links in the simulated gridlock area were investigated for the development of the identification of bottlenecks and the prediction of gridlock.

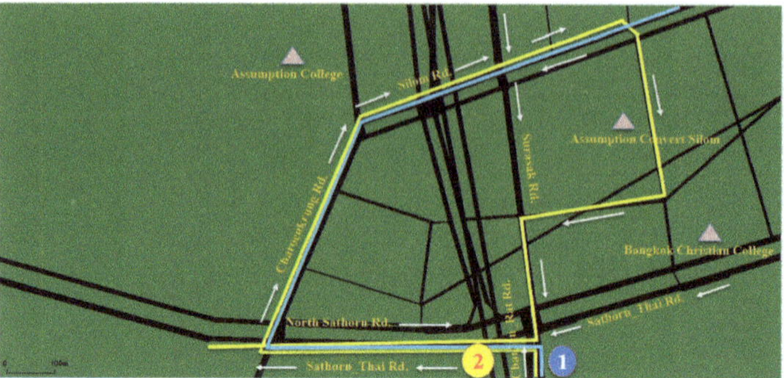

Figure 3. Simulated additional traffic routes representing parent trips to drop off students at local schools during morning rush hour.

Figure 4. Links in the Sathorn critical region.

4. Methodology

4.1. Traffic Data Processing

A traffic network is comprised of a set of links and a set of junctions or intersections in the area of a typical urban transportation network. The network-wide congestion state of links is expressed as a 2×2 matrix indexed in time (the number of time steps T) and space (the number of links N). Given the intersection and link topological characteristics, the intent is to develop a bottleneck identification for each intersection in the network using GPS vehicles as the input data. The initial step is to identify the traffic congestion state of each link using GPS data. The GPS data are collected with varying time resolutions (1 s, 5 s, 10 s, 15 s, 20 s, 25 s, 30 s, 35 s, 40 s, 45 s, 50 s, 55 s, 60 s) to represent the congestion state of each upstream link u and downstream link d, for each of the intersection areas. Herein, let e denote an arbitrary link index for upstream link u or downstream link d. The mean link speed V_e^t of link e at time t is computed from the arithmetic mean of the speed values of all GPS vehicles on that link e at time t. In the matrix, the mean link speed V_e^t values for all links e and time t is expressed as:

$$\begin{bmatrix} V_1^1 & V_1^2 & \cdots & V_1^T \\ V_2^1 & V_2^2 & \cdots & V_2^T \\ \vdots & \vdots & \ddots & \vdots \\ V_N^1 & V_N^2 & \cdots & V_N^T \end{bmatrix}$$

4.2. Description of Bottleneck and Gridlock Identification

In this study, the congestion state for each upstream link u and downstream link d is defined in terms of the low mean link speed indicator C_e^t at time t of an arbitrary link e, as shown in Equation (1).

The traffic congestion is set with the cut-off mean link speed threshold (5 km/h). If the mean link speed is less than or equal to the threshold, then the low mean link speed indicator C_e^t of link e at time t is 1 (congested); otherwise, it is 0 (not congested).

$$C_e^t = \begin{cases} 1, & \text{when } V_e^t <= 5 \ km/h \\ 0, & \text{otherwise} \end{cases} \tag{1}$$

However, the low mean link speed for both the upstream and downstream can be affected by the red traffic signal light. To filter out such a red traffic signal light effect, therefore, if the low mean link speed indicator C_e^t alerts that there is congestion for a long enough time interval of length w, then this congestion state is said to be persistent for the corresponding link. For instance, the time interval w is 10 min herein, which is long enough to filter out the usual red time intervals in practical scenarios in the Chula-SSS dataset [7]. The persistently low mean link speed indicator $\tilde{C}_e^t$ of link e at time t can then be expressed as:

$$\tilde{C}_e^t = \prod_{\tau=0}^{w-1} C_e^{t-\tau} \tag{2}$$

A bottleneck occurrence at an intersection can be determined by considering two locations, upstream link u and downstream link d at the intersection. This study considers the direction of the upstream and downstream only on the conflicting approaches in the loop, as shown in Figure 1. A bottleneck is said to occur if and only if the low mean link speed indicator for any possible pair of upstream and downstream links at that intersection drops below a cut-off speed threshold. Let P be the number of upstream and downstream link pairs at intersection I, and (u_j, d_j) denotes the jth pair of upstream and downstream links, $(j = 1, 2, \ldots, P)$. For intersection I, define the bottleneck indicator B_I^t at time t as Equation (4) based on the low mean link speed indicators C_u^t and C_d^t for any upstream and downstream link pair (u, d) in I. If all pairs of upstream and downstream links of the intersection I are congested at t, then we say that there is a bottleneck in this intersection I at time t.

$$C_{(u,d)}^t = C_u^t \cdot C_d^t \tag{3}$$

$$B_I^t = \prod_{\forall (u,d) \in \{(u_1,d_1),(u_2,d_2),\ldots,(u_P,d_P)\}} C_{(u,d)}^t \tag{4}$$

A temporary low-speed intersection does not necessarily represent a bottleneck, but an active bottleneck always results in a low-speed intersection. Therefore, the persistent bottleneck indicator $\tilde{B}_I^t$ is used to confirm an active bottleneck for every time step τ within a time interval of length w, as shown in Equation (6).

$$\tilde{C}_{(u,d)}^t = \tilde{C}_u^t \cdot \tilde{C}_d^t \tag{5}$$

$$\tilde{B}_I^t = \prod_{\forall (u,d) \in \{(u_1,d_1),(u_2,d_2),\ldots,(u_P,d_P)\}} \prod_{\tau=0}^{w-1} \tilde{C}_{(u,d)}^{t-\tau} \tag{6}$$

$$G_L^t = \sum_{I \in L} B_I^t \tag{7}$$

$$\tilde{G}_L^t = \sum_{I \in L} \tilde{B}_I^t \tag{8}$$

With the bottleneck intersection identification, gridlock can be defined as the severe congestion state of a critical region that forms a loop of adjacent intersections in an urban road network. Let L be a set of potentially gridlock-looped intersections and $L = \{I_1, I_2, \ldots, I_J\}$. For the Chula-SSS dataset, a recurrent gridlock loop is shown in Figure 5 with intersections within yellow dotted circles. The gridlock indicator for loop L, G_L^t, is triggered at time t if the bottleneck indicator B_I^t for every intersection $I \in L$ at time t is triggered, as shown in Equation (7). Consequently, the persistent gridlock is triggered due to the active bottleneck of all intersections, as shown in Equation (8). The computed gridlock labels quantizing the overall effect of the bottleneck indicators $\tilde{B}_I^t$ in loop L for all intersections are as shown in Figure 5. The computed gridlock labels are 0, 1, 2, 3, 4, and 5 based on the percentage of congested intersections in loop L. For example, if all intersections are not congested, then the computed gridlock label is 0; if 20% of intersections are congested, then the computed gridlock label is 1; if 40% of intersections are congested, then the computed gridlock label is 2, and so on, as shown in Table 1.

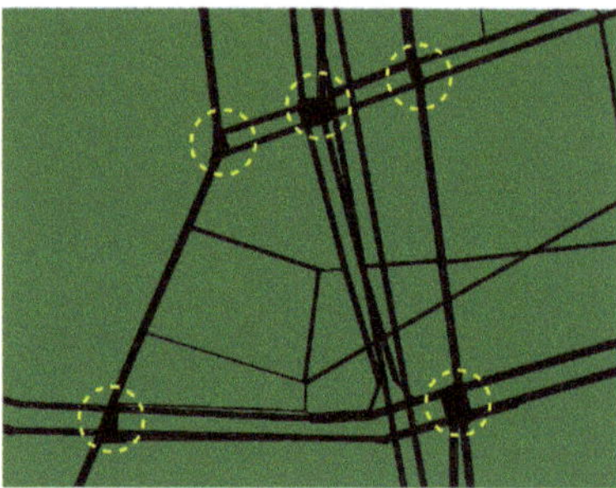

Figure 5. Gridlock-looped intersections in the Sathorn area.

Table 1. Gridlock labels.

Bottleneck Intersections	Gridlock Labels
all intersections are congested	5
80% of intersections are congested	4
60% of intersections are congested	3
40% of intersections are congested	2
20% of intersections are congested	1
all intersections are not congested	0

4.3. Effect of Sample Size on Gridlock Detection

The sample size selection is fundamental to any traffic engineering analysis, including bottleneck and gridlock analysis. Firstly, this section attempts to determine the time resolution based on the vehicles' mean speed on the links in the interested area. The GPS data collection is simulated in the Chula-SSS dataset with different time resolutions (1 s, 5 s, 10 s, 15 s, 20 s, 25 s, 30 s, 35 s, 40 s, 45 s, 50 s, 55 s, 60 s) for the simulation time interval of three hours from 6 am to 9 am. Figure 6 shows the mean speed of all GPS vehicles on the upstream links (Sathorn_Thai_1, Charoen_Rat, Surasak) and downstream link (Sathorn_Thai_2) at the Surasak intersection. In this Figure 6, the mean speed of each upstream and downstream link reaches the congested conditions after the low mean speed. The standard deviation is the measure of the variability of the data distribution. Herein, the standard deviation of the mean speed of GPS vehicles on each upstream/downstream link for the sample size of 30% in each time resolution is shown in Figure 7. The reported results show that the standard deviation of the mean speed of all GPS vehicles on each upstream/downstream link obtained from 1 s at a high resolution is not statistically different from that obtained from 60 s at a low resolution. This suggests that any time resolution in the range of 1 s to 60 s can be used for bottleneck and gridlock analysis. Therefore, this study selected, for subsequent experimental investigations, the value of 60 s, which is currently available using GPS vehicles in practice for Bangkok's targeted area in this research study.

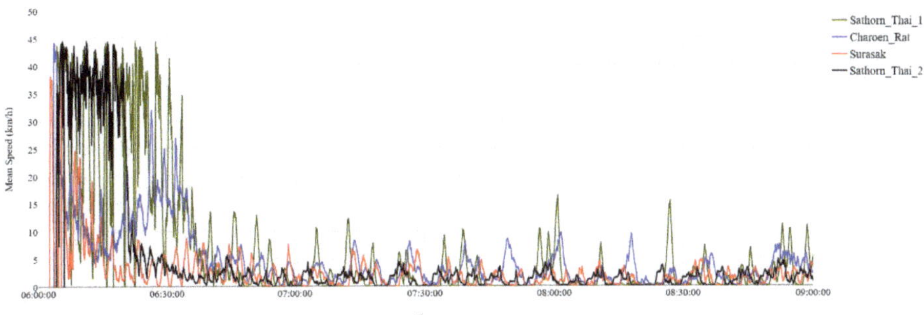

Figure 6. Mean speed of the upstream links (Sathorn_Thai_1, Charoen_Rat, Surasak) and downstream link (Sathorn_Thai_2) at the Surasak intersection.

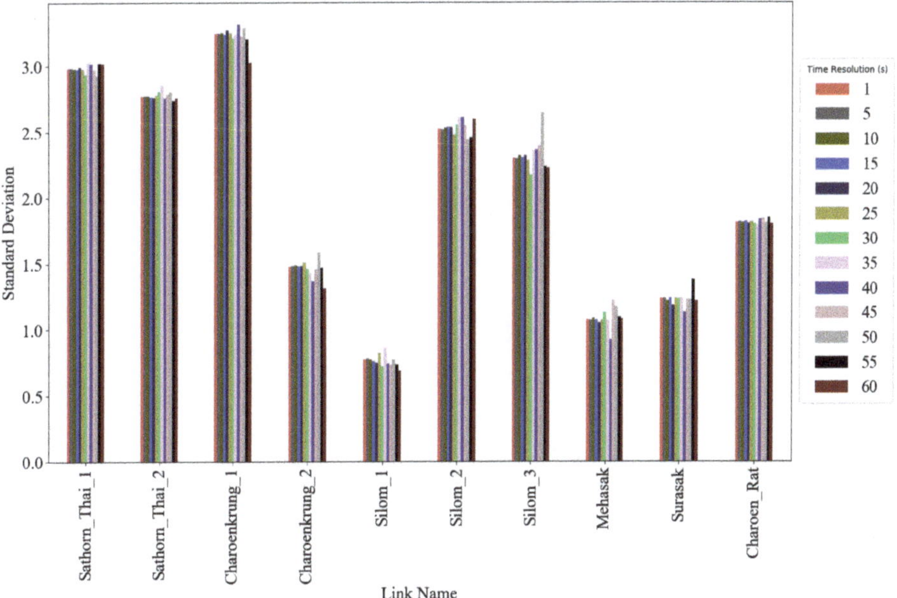

Figure 7. Standard deviation of the mean speed of GPS vehicles on each upstream/downstream link for a sample size of 30% for each time resolution.

Besides, this study investigated the sample size with which to experiment to analyze the status of gridlock with the multiple categories of gridlock labels based on bottleneck intersections. These gridlock labels are based on the identification of gridlock, as defined in Section 4.2. In particular, from the simulated movements of all GPS vehicles, we can compute the Detection Rate (DR) and the False Alarm Rate (FAR) of detected gridlock labels. DR and FAR are based on the True Positive (TP), True Negative (TN), False Positive (FP), and False Negative (FN) by:

$$DR = \left(\frac{TP}{TP + FN} \right) \tag{9}$$

$$FAR = \left(\frac{TN}{TN + FP} \right) \tag{10}$$

where TP, TN, FP, and FN measure the number of time steps for which corresponding results occur for the detection of gridlock label values. The detection of a gridlock label is said to be positive if, and only if, Equation (8) returns that gridlock label as the output. Otherwise, the detection is said to be negative for that gridlock label. The actual value of the gridlock label is computed by Equation (8) when 100% of simulated vehicles are assumed to have GPS data. In the case of less than 100% of available GPS data, the positive (or negative) detection results are said to be true or false according to whether or not the detection results are the same as those obtainable by the actual value of the gridlock label.

This section explores the effect of sample size on the identification of gridlock labels to understand the detection rate and false alarm rate. The detection rate and false alarm rate resulting from the settings of different sample percentages of GPS vehicles' penetration ratio (1%, 5%, 10%, 15%, 20%, 25%, 30%, 35%, 40%, 45%, 50%) for the identification of gridlock labels are plotted in Figures 8 and 9. Here, DR_1, DR_2, DR_3, DR_4, and DR_5 are indicated for the detection rate of the gridlock labels 1, 2, 3, 4, and 5. Furthermore, FAR_1, FAR_2, FAR_3, FAR_4, and FAR_5 are indicated for the false alarm rate of the gridlock labels 1, 2, 3, 4, and 5. These diagrams depict the detection rate and false alarm rate in each sample size with the time resolution of 60 s. In general, DR denotes the percentage of correctly detected gridlock. The algorithm works well in identifying gridlock labels if the DR is close to 100%. Notably, increasing the number of GPS vehicles will dramatically affect the detection rate and false alarm rate.

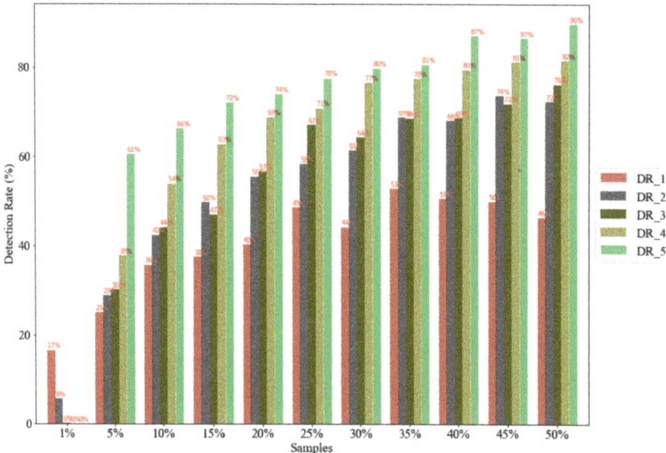

Figure 8. Detection Rate (DR) for all gridlock labels.

The data coverage of the road network is helpful for bottleneck and gridlock analysis. The higher the penetration ratio of vehicles with GPS in the road network is, the better the data coverage capability, and consequently, the detection rate increases and false alarm rate decreases, as shown in Figures 8 and 9 for each gridlock label, except the sample size of 1% for low data coverage. Furthermore, it should be noted that the identification is most accurate for the gridlock label 5 (the network has been locked) in terms of both DR and FAR because the traffic flow of the gridlock-lopped intersections reaches almost the halted state. With the same logic for the explanation, on the contrary, identifying low gridlock labels (where the network is lightly or moderately loaded) is less accurate. However, such identification for lower gridlock labels poses less concern than when the whole network is gridlocked. Thus, a significant benefit of studying the detection rate and false alarm rate is that they allow for a means of gridlock prediction under the circumstances of the GPS vehicle penetration ratio. At the gridlock label 5, which is the most critical one for gridlock management, it is finally noted that to achieve the practical target of at least 80% DR and at most 10% FAR, Figures 8 and 9 indicate that 30% of GPS vehicles are needed.

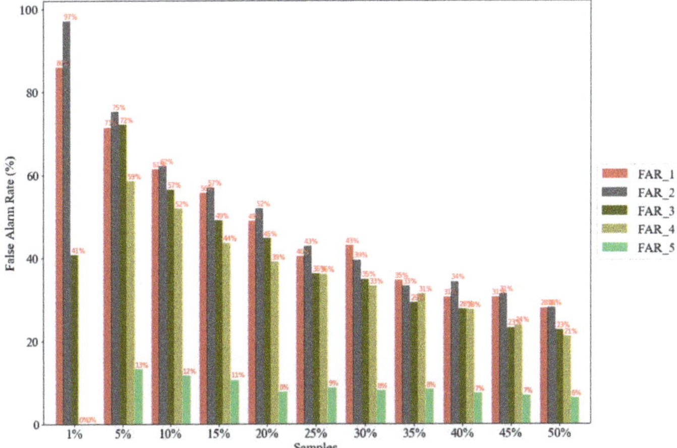

Figure 9. False alarm rate for all gridlock labels.

5. Urban Gridlock Prediction Based on Intersection Bottlenecks with Long Short-Term Memory

Time series modeling and prediction have been an active area of research due to the wide variety of applications [32]. Deep learning has been applied in time series analysis for both classification and prediction [33]. Prediction for time series is usually based on historical measurements. The better the learning of the inter-dependencies among past observations is modeled, the more accurate the prediction can be. Long Short-Term Memory (LSTM) [23] has been developed as an extension of the recurrent neural network with the advantages of learning long-range time dependencies.

5.1. Time Series Construction for Gridlock Prediction

The time series prediction methods have been developed over the years, advancing from simple regression techniques to statistical and intelligent algorithms. Short-term traffic flow prediction is a time series analysis that aims to predict traffic flow for a particular time based on historical traffic flow, as shown in Figure 10. This research adopted this concept to predict the severe road traffic congestion based on the historical time series congestion status of the intersections and gridlocks in the urban road network.

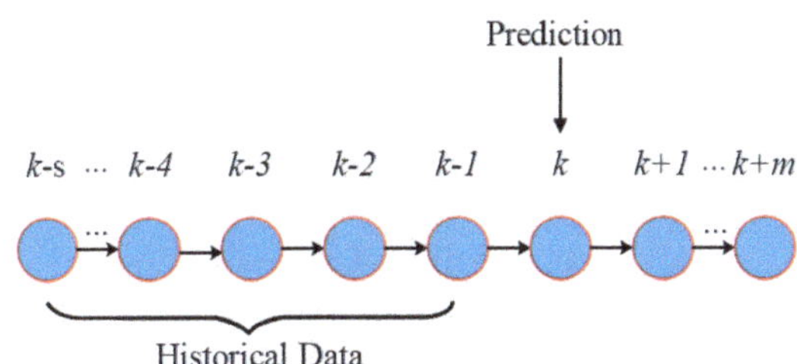

Figure 10. Time series prediction.

In our formulation, firstly, the time series observations of bottleneck indicators $B^t_{I_1}, ..., B^t_{I_J}$ and computed gridlock indicator G^t_L for each t are calculated. The values of all such indicators are then transformed to a multivariate time series of persistent bottleneck indicators for all intersections $\tilde{B}^t_{I_1}, ..., \tilde{B}^t_{I_J}$ and computed persistent gridlock indicator $\tilde{G}^t_L$, as defined earlier in Section 4.2. A multivariate time series is denoted as $X = \left\langle \tilde{B}^T_{I_1}, ..., \tilde{B}^T_{I_J}, \tilde{G}^T_L \right\rangle$ with time stamp T. To take into account all the intersections that can form a potential gridlock loop, define x^t as the vector of all persistent bottleneck indicators of $\tilde{B}^t_{I_1}, \ldots, \tilde{B}^t_{I_J}$ and persistent gridlock indicator $\tilde{G}^t_L$ for each t, as shown in Table 2. As indicated in Figure 10, to distinguish the time instances used in simulating detailed system dynamics and in obtaining the time series observations with the potentially more extended time sampling period, we change the previously used time indices from t to k for the following gridlock prediction formulation.

Table 2. Observation instances of observed and computed time series.

Instance	Bottleneck					Gridlock
x^1	$\tilde{B}^1_{I_1}$	$\tilde{B}^1_{I_2}$	.	.	$\tilde{B}^1_{I_J}$	$\tilde{G}^1_L$
x^2	$\tilde{B}^2_{I_1}$	$\tilde{B}^2_{I_2}$	.	.	$\tilde{B}^2_{I_J}$	$\tilde{G}^2_L$
x^3	$\tilde{B}^3_{I_1}$	$\tilde{B}^3_{I_2}$	.	.	$\tilde{B}^3_{I_J}$	$\tilde{G}^3_L$
$\vdots$	$\vdots$	$\vdots$	$\vdots$	$\vdots$	$\vdots$	$\vdots$
x^T	$\tilde{B}^T_{I_1}$	$\tilde{B}^T_{I_2}$	.	.	$\tilde{B}^T_{I_J}$	$\tilde{G}^T_L$

Define the k-th observation value as the original time series value sampled at time $t = kT_s$ where T_s denotes the length of the time sampling interval. The basic idea in this study is to predict the k-th observation y^k fed to produce $\tilde{G}^{kT_s}_L$ from a finite set of s time-lagged observations; we aim at predicting in a rolling forecasting fashion, herein defined as:

$$z^k = \{x^{(k-1)T_s}, \ldots, x^{(k-s)T_s}\} \tag{11}$$

where $x^{(k-1)T_s}, \ldots, x^{(k-s)T_s}$ denote the corresponding time sampled values for the k-th observation. The input historical bottleneck and computed gridlock sequence are used to predict gridlock at the k-th observation. Figure 11 shows the unrolled network of time-lagged observation to predict y^k over the set of time-lagged observations z^k. In the experiments, we consider $z^k = x^{(k-1)T_s}$, i.e., $s = 1$ for the case of using one time-lagged observation, as well as trying to increase s, the number of time-lagged observations.

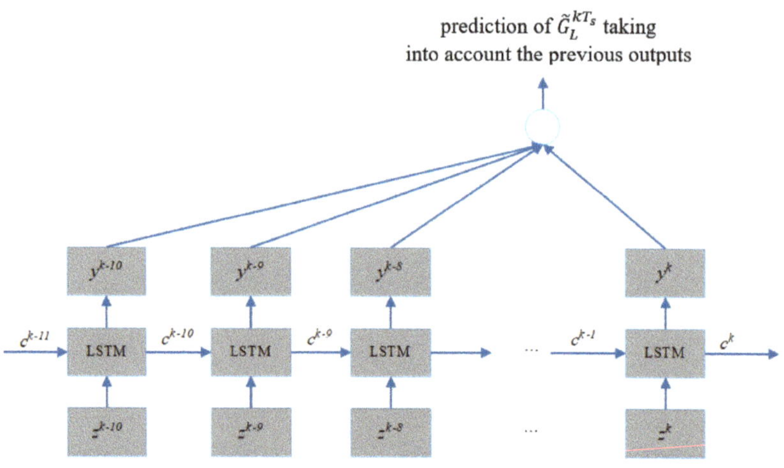

Figure 11. Unrolled LSTM network.

5.2. Formulation for the Prediction of the Time Sampled Persistent Gridlock Indicator

The LSTM has the advantage of capturing temporal information and is approved to be adopted in time series modeling. It is a special kind of recurrent neural network with a purpose-built memory cell representing long-term dependencies in the time series data.

The memory cell has three units, namely the input gate (i^k), the forget gate (f^k), and the output gate (o^k), to maintain the information of a long period. The forget gate determines what information is required to be rejected from the input z^k. The input gate i^k determines what information z^k is required to be kept, and the output gate o^k produces the output based on z^k. They are composed of a sigmoid neural network layer (σ), pointwise multiplication operation, and tanh function, as depicted in Figure 12. w_f, b_f, w_i, b_i, w_o, and b_o are the weight matrices and bias vectors of each gate. The gates are updated in LSTM by [23]:

$$f^k = \sigma(w_f \cdot [z^k, y^{k-1}] + b_f)$$
$$i^k = \sigma(w_i \cdot [z^k, y^{k-1}] + b_i) \tag{12}$$
$$o^k = \sigma(w_o \cdot [z^k, y^{k-1}] + b_o)$$

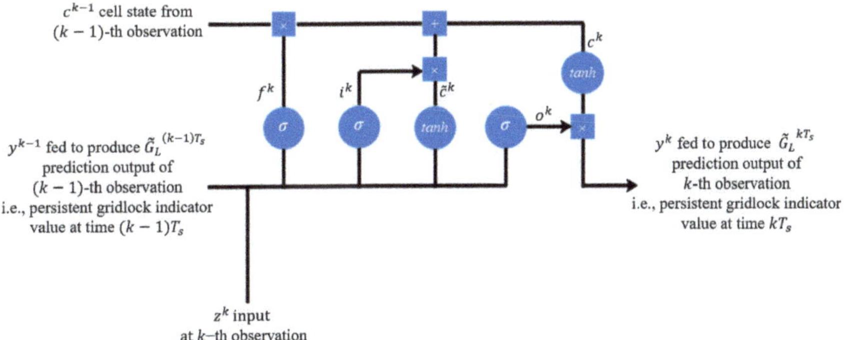

Figure 12. LSTM memory cell for time series prediction of the persistent gridlock indicator with the time sampling interval of T_s.

After updating the gates, the updated values from the input and forget gates are used for updating the current state c^k of the LSTM unit based on $\tilde{c}^k$ and $c^{k-1} \cdot w_c$ and b_c are the weight metrics and bias vector, and tanh is the activation function. The output y^k is triggered by the output gate o^k as defined in Equation (13), which yields the activated output to the next LSTM unit.

$$\tilde{c}^k = \tanh(w_c \cdot [z^k, y^{k-1}] + b_c)$$
$$c^k = f^k \cdot c^{k-1} + i^k \cdot \tilde{c}^k \tag{13}$$
$$y^k = o^k \cdot \tanh(c^k)$$

Additionally, the activation sigmoid function σ is used to map the range of [0,1], and the tanh function is applied to map the range of [−1,1].

6. Evaluation of the LSTM Based Prediction Experiment on the Chula-SSS Dataset

This study used a dataset that has been collected from the GPS vehicles sampled from the calibrated Chula-SSS dataset in SUMO using different random seed numbers for a total of 100 simulated days (each from 6 am to 9 am). The GPS data collection was periodically sampled every 60 s during the simulation time interval. To develop and evaluate the LSTM model, the dataset was divided into training data containing 80 simulated days and testing data containing the remaining 20 simulated days from a total of 100 simulated days. Each simulated day has 181 time steps (every 60 s within

6 am to 9 am) and 6 features by combining the 5 persistent bottleneck indicators of 5 intersections and 1 persistent gridlock indicator. The important step is to select the model hyperparameters. In this investigation, the hyperparameters of LSTM include the number of neurons to improve the network's learning capacity, the number of epochs, and the batch size, as shown in Table 3. The number of epochs defines the number of times that the learning algorithm will work through the all training data. The models were developed with the standard Keras library [34]. Additionally, the standard gradient descent algorithm, ADAM [35], was used to update the weight values given the batch size value.

Table 3. Description of the LSTM model setup.

Item	Description
Prediction Target	Gridlock labels (0, 1, 2, 3, 4, 5)
Input Variable	Multivariate time series observations for all intersections
Training Parameters	epochs, number of neurons, batch size

To evaluate the effectiveness of the gridlock prediction model, we calculate the Root Mean Squared Error (RMSE) and Mean Absolute Error (MAE) [36] in estimating the time sampled values of the persistent gridlock indicator. These metrics are calculated by comparing the time series's target values and the corresponding time series for the concatenation of all the testing data, each lasting for three hours. The proposed approach for predicting time series is applied to predict gridlock based on the bottleneck state for each intersection in the experiments with different sample sizes by varying the GPS vehicle penetration ratio in percentage and with various settings on the time-lagged observation period (sT_s) and the time sampling interval (T_s). Figures 13–15 report the obtained results, i.e., the RMSE and the MAE, as well as the required computation time of LSTM.

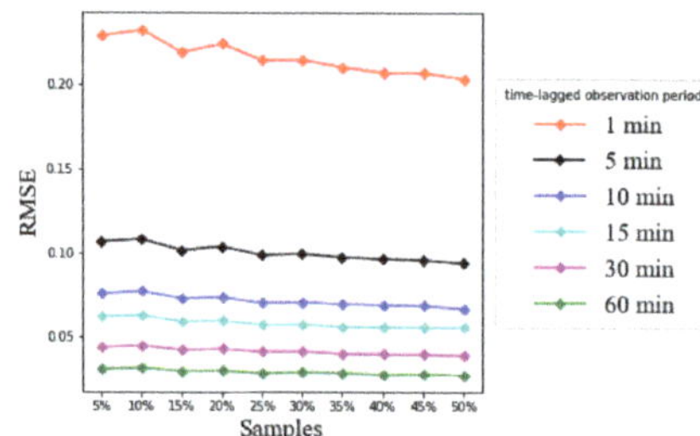

Figure 13. RMSE with varied time-lagged observation period.

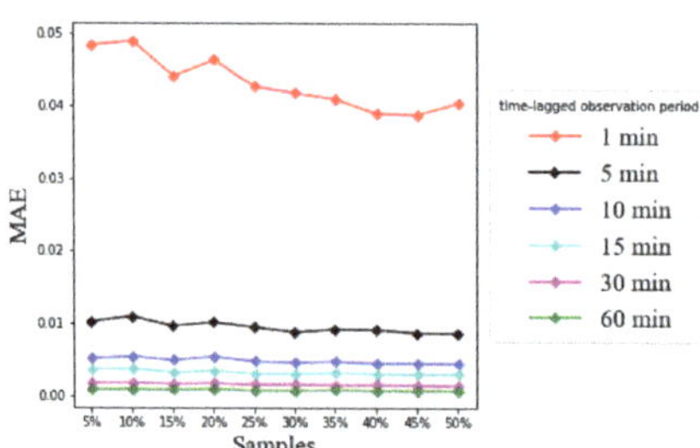

Figure 14. MAE with varied time-lagged observation period.

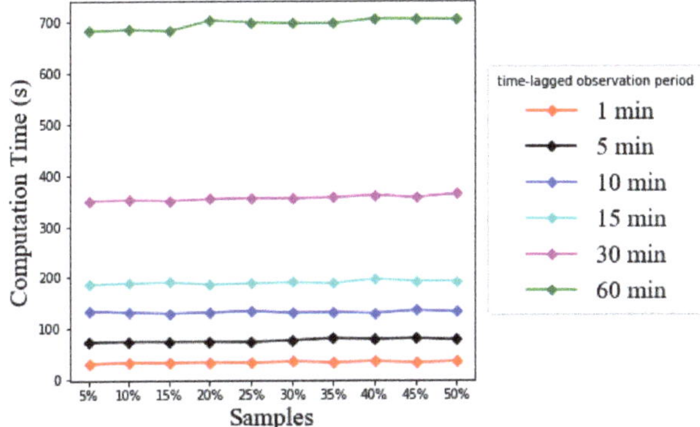

Figure 15. Computation time with varied time-lagged observation period.

6.1. Effect of Time-Lagged Observation Period

Firstly, the time-lagged observation period (sT_s) was varied as 1, 5, 10, 15, 30, and 60 min, as shown in Figures 13–15. Herein, the time sampling interval (T_s) for every 60 s was used. Figure 13 shows the RMSE values for different time-lagged observation periods using 10 neurons of LSTM outputs linearly combined with the final dense layer with one prediction output, 10 epochs, and a batch size of 50. Figure 14 shows the MAE values for different time-lagged observation periods using the same parameters. These hyperparameter values were searched by brute force and finally selected to show the achieved performance of LSTM in our considered case studies. To compare the prediction error values, this study considered as a baseline the lowest 1 min time-lagged observation period, i.e., LSTM is allowed to look back into the past for only 1 min. Figures 13 and 14 show that the RMSE and MAE eventually decrease with the increase of the past observation interval length allowable for LSTM based prediction. When the time-lagged observation is 1 min, there is a big gap between the RMSE values and the other time-lagged observations of 5 min to 60 min. It is understandable that the short historical observation interval cannot provide adequate information for accurate prediction. However, the gap of the RMSE and MAE values between 5 min and other time-lagged observation periods from 10 min to 60 min is relatively not as much as that from 1 min to 5 min.

Although a short historical observation interval does not generally provide accurate prediction, Figures 13 and 14 suggest that the RMSE and MAE values are still acceptable even at the 1 min time-lagged observation, i.e., less than 0.25 for RMSE and 0.05 for MAE from the presumed linear scale of five. On the other hand, if LSTM has more information on the extended time-lagged observation, the RMSE value is reducible to as low as 0.02, and the MAE value is as low as 0.001 with the 60 min time-lagged observation. Therefore, the result confirms that LSTM has good performance over the whole practical range of time-lagged observations. The performance improves further when the input time-lagged observations are sufficiently long.

Although prediction accuracy is improved, if the number of time-lagged observations is increased, the computation is unavoidably increased, as shown in Figure 15. The computation time reported here is based on the currently used hardware of NVIDIA GeForce GTX 1080 Ti with 11 GB memory. At the 60 min time-lagged observation period, it is found that the computation time is up to 700 s or almost 12 min. Such a computation time can be further reduced if the hardware capacity is upgraded. In practice, engineers can design the needed hardware so that the computation can be completed within the time sampling interval, resulting in a real-time prediction effect.

Figure 16 shows the predicted and actual gridlock labels with 1 min time-lagged observation using the 5% sample size. Figure 17 shows the detection rate for all gridlock labels after prediction with LSTM. In Section 4.3, this study presents the effect of sample size to detect gridlock, and the detection rate is 80% on the sample size of 30% GPS vehicles. However, after prediction with LSTM, our proposed LSTM model can predict more gridlock labels, 1, 2, and 3, as low as the 1% sample, but the 1% sample cannot predict the gridlock labels 4 and 5 using the insufficient information of the loop. In Figure 17, the reported DR is 93% for the 5% sample size. This reported detection rate after LSTM confirms that LSTM has good performance using the time-lagged observation in the rolling forecast fashion.

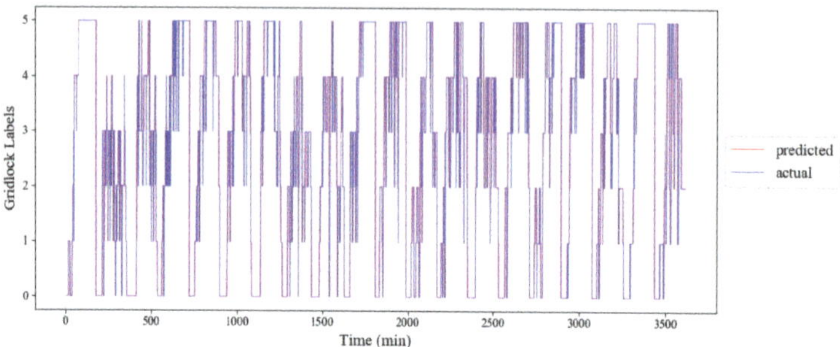

Figure 16. Example of actual and predicted gridlock labels with the 1 min time-lagged observation using the 5% sample size.

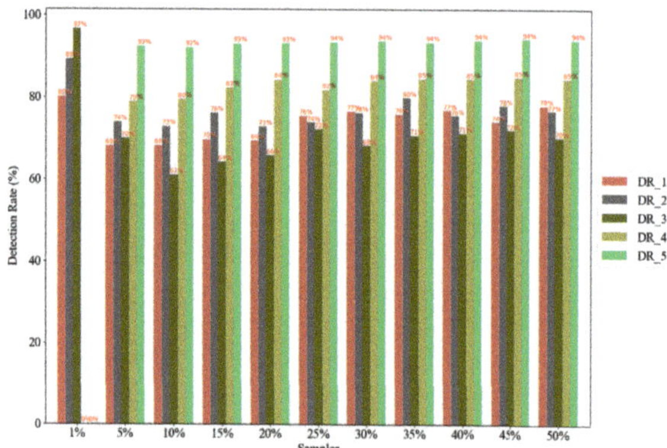

Figure 17. Detection rate for all gridlock labels after prediction with LSTM.

In Figures 18 and 19, the reported RMSE and MAE values are decreased to as low as 0.03 and 0.001 with the 60 min time-lagged observation using the 3% sample. Therefore, the results confirm that the required minimum sample size of 3% GPS vehicles traveling on each link of the loop is enough to predict the gridlock. Using the mean speed of this penetration ratio of GPS vehicles, in practice, can successfully and effectively detect gridlock.

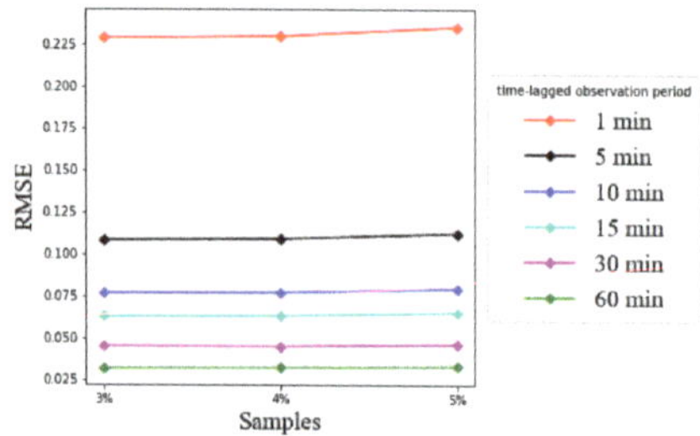

Figure 18. RMSE of sample sizes 3% to 5% with varied time-lagged observation period.

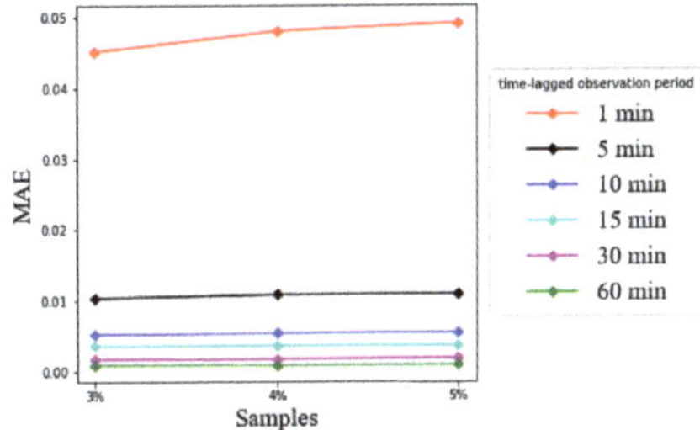

Figure 19. MAE of sample sizes 3% to 5% with varied time-lagged observation period.

6.2. Effect of Time-Sampling

A short time sampling interval of data collection refers to a high sampling rate, while a long time sampling interval of data collection refers to a low sampling rate. The higher sampling rate of GPS vehicle trajectories gives more sampling points to form the trajectory path. However, because of the strict limits of battery storage for hand-held GPS-embedded devices carried inside moving vehicles, they are often unable to collect high sampling rate data for a long period [37]. Therefore, the time sampling rate interval of GPS vehicle data is an important issue. To study the effect of the sampling interval in gridlock prediction, this study used the time sampling interval T_s of the GPS vehicle data in Figures 20–22 at 5, 10, and 15 min. Like in Figures 13–15, we selected the LSTM hyperparameter values of 10 neurons, 10 epochs, and a batch size of 50.

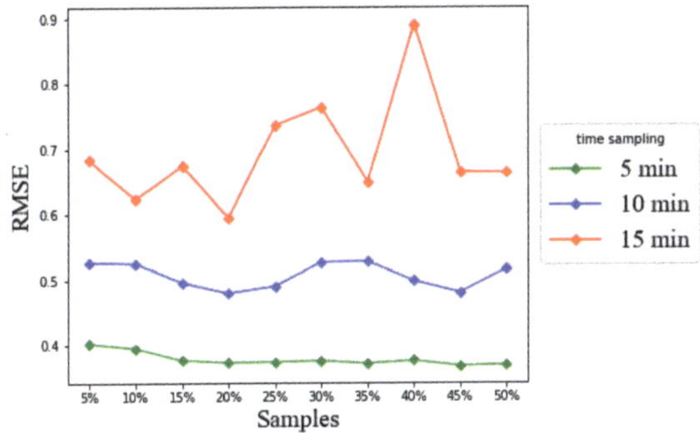

Figure 20. RMSE with varied time sampling.

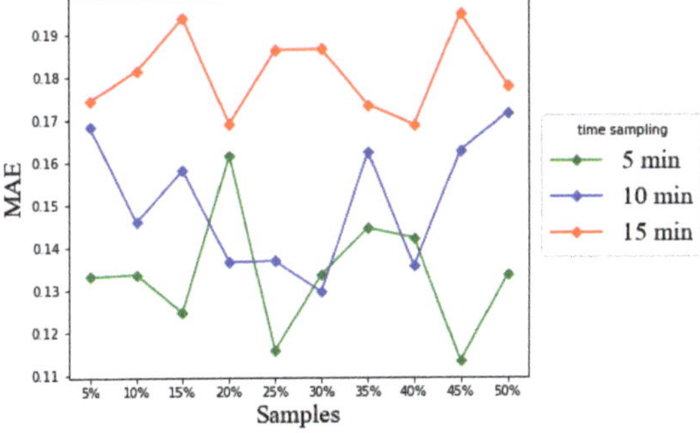

Figure 21. MAE with varied time sampling.

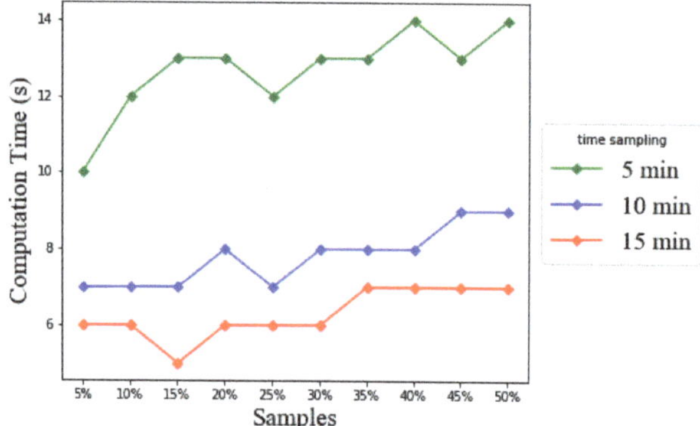

Figure 22. Computation time with varied time sampling.

Our findings suggest that the time sampling interval directly affects the achievable accuracy of LSTM prediction for persistent gridlock indicators. In Figure 20, the reported prediction error increases as the time sampling interval increases. The prediction error obtained from the time sampling interval of 5 min is at most 0.4 for RMSE for the minimally required 5% GPS vehicle sample. However, a long time sampling interval results in a large RMSE value, for example up to 0.9 for RMSE for the 15 min time sampling. From the computation time point of view, the longer the time sampling interval, the less the computation time, as shown in Figure 22. For example, the 15 min time sampling lasts only around 6 s or 7 s.

In practice, one can evaluate the quantized gridlock label by rounding off the predicted gridlock label value first. In this case, an RMSE less than 0.5 should be considered as acceptable. Under such practical concerns, based on Figures 20–22, the proposed LSTM based model can predict the gridlock label for effectively 5 min into the future and at the affordable up to 15 s computation time. That lead time of 5 min is comparable to a reasonable cycle length of adjusting traffic signal lights. Therefore, by foreseeing such gridlock occurrences by one control cycle, at least, the operational traffic signal controller can adaptively try to correct its signal interval settings to either mitigate the occurring gridlock effects or prevent the gridlock from happening in the first place.

7. Conclusions

This study proposes bottleneck based gridlock prediction on simulated GPS vehicles using spatiotemporal time series analysis based LSTM. The investigation shows that LSTM can predict the gridlocks with a long range of time dependencies by conducting the experiments on different time-lagged observation periods and time sampling. The sample size of 30% GPS vehicles is needed for geographic data coverage of bottleneck and gridlock analysis before using LSTM. The RMSE result of 0.02 with the 60 min time-lagged observation confirms that LSTM prediction with current hardware capacity has satisfying performance if the input time-lagged observations are sufficiently long. The time sampling interval of 5 min reports 0.4 for RMSE with the required sample of 5% GPS vehicles. The reported results suggest the effect of time sampling using the data collection interval of 5 min. Our proposed LSTM model can predict the gridlock labels for effectively 5 min into the future with a 15 s computation time. Therefore, if our LSTM model has information on the previous time-lagged observation of a 5 min data collection interval, it can predict the future 5 min of the gridlock occurrences. The traffic signal controller can adapt to correct the signal timings to alleviate the gridlock effects in the loop. The reported RMSE and MAE values are decreased to as low as 0.03 and 0.001, with the 60 min time-lagged observation using the 3% sample with LSTM. Therefore, the results confirm that the required minimum sample size of 3% GPS vehicles traveling on each link of the loop is enough to predict the gridlock. Using the mean speed of this penetration ratio of GPS vehicles, in practice, can successfully and effectively detect the gridlock. The reported results suggest that the percentage of simulated GPS vehicles using different random seed numbers can give the possibility of bottleneck and gridlock identification, as well as gridlock prediction using LSTM.

Author Contributions: Conceptualization, E.E.M., H.O., C.S., and C.A.; data curation, E.E.M.; formal analysis, E.E.M., H.O., and C.A.; funding acquisition, C.A.; investigation, E.E.M., H.O., and C.A.; methodology, E.E.M. and C.A.; resources, H.O., C.S., and C.A.; software, E.E.M.; supervision, H.O., C.S., and C.A.; validation, E.E.M. and C.A.; visualization, E.E.M.; writing, original draft, E.E.M.; writing, review and editing, H.O., C.S., and C.A. All authors have read and agreed to the published version of the manuscript.

Funding: This work was supported in part by the Collaborative Research Project entitled Road Traffic Monitoring and Prediction System to provide Intelligent Transportation System, in part by the JICA Project for AUN/SEED-Net, Japan, and in part by the Asi@Connect's Data-Centric IoT-Cloud Service Platform for Smart Communities (IoTcloudServe@TEIN) project with Grant Contract ACA 2016/376-562 under the umbrella of the Smart-Mobility@Chula demonstration site.

Acknowledgments: The authors gratefully acknowledge the AUN/SEED-Net Scholarship and all members of the smart mobility project at the Department of Electrical Engineering, Chulalongkorn University, Bangkok, Thailand.

Conflicts of Interest: The authors declare no conflict of interest.

References

1. Qi, H.; Chen, M.; Wang, D. Recurrent and non-recurrent bottleneck analysis based on traffic state rank distribution. *Transp. B Transp. Dyn.* **2019**, *7*, 275–294. [CrossRef]
2. Sun, H.; Wu, J.; Ma, D.; Long, J. Spatial distribution complexities of traffic congestion and bottlenecks in different network topologies. *Appl. Math. Model.* **2014**, *38*, 496–505. [CrossRef]
3. Mon, E.E.; Ochiai, H.; Saivichit, C.; Aswakul, C. Recurrent and Non-recurrent Congestion Based Gridlock Detection on Chula-SSS Urban Road Network. *EPiC Ser. Comput.* **2019**, *62*, 158–171.
4. Guo, J.; Huang, W.; Williams, B.M. Adaptive Kalman filter approach for stochastic short-term traffic flow rate prediction and uncertainty quantification. *Transp. Res. Part C Emerg. Technol.* **2014**, *43*, 50–64. [CrossRef]
5. Gong, L.; Fan, W. Developing a systematic method for identifying and ranking freeway bottlenecks using vehicle probe data. *J. Transp. Eng. Part A Syst.* **2018**, *144*, 04017083. [CrossRef]
6. Yun, M.; Qin, W. Minimum Sampling Size of Floating Cars for Urban Link Travel Time Distribution Estimation. *Transp. Res. Rec.* **2019**, *2673*, 24–43. [CrossRef]
7. Aswakul, C.; Watarakitpaisarn, S.; Komolkiti, P.; Krisanachantara, C.; Techakittiroj, K. Chula-SSS: Developmental Framework for Signal Actuated Logics on SUMO Platform in Over-saturated Sathorn Road Network Scenario. In *SUMO 2018-Simulating Autonomous and Intermodal Transport Systems*; EPiC Series in Engineering; EasyChair: Berlin, Germany, 2018; Volume 2, pp. 67–81. [CrossRef]
8. Lee, W.H.; Tseng, S.S.; Shieh, J.L.; Chen, H.H. Discovering traffic bottlenecks in an urban network by spatiotemporal data mining on location-based services. *IEEE Trans. Intell. Transp. Syst.* **2011**, *12*, 1047–1056. [CrossRef]
9. Wu, X.; Liu, H.X.; Gettman, D. Identification of oversaturated intersections using high-resolution traffic signal data. *Transp. Res. Part C Emerg. Technol.* **2010**, *18*, 626–638. [CrossRef]
10. Yue, W.; Li, C.; Mao, G. Urban Traffic Bottleneck Identification Based on Congestion Propagation. In Proceedings of the 2018 IEEE International Conference on Communications (ICC), Kansas City, MO, USA, 20–24 May 2018; pp. 1–6.
11. Tang, L.; Wang, Y.; Zhang, X. Identifying Recurring Bottlenecks on Urban Expressway Using a Fusion Method Based on Loop Detector Data. *Math. Probl. Eng.* **2019**, *2019*, 5861414 . [CrossRef]
12. Zhao, W.; McCormack, E.; Dailey, D.J.; Scharnhorst, E. Using truck probe GPS data to identify and rank roadway bottlenecks. *J. Transp. Eng.* **2013**, *139*, 1–7. [CrossRef]
13. Jimenez, A.L.; Rodriguez-Valencia, A. Exploratory Methodology for Identification of Urban Bottlenecks Using GPS Data. *J. Traffic Transp. Eng.* **2016**, *4*, 280–290.
14. Jose, R.; Mitra, S. Identifying and Classifying Highway Bottlenecks Based on Spatial and Temporal Variation of Speed. *J. Transp. Eng. Part A Syst.* **2018**, *144*, 04018075. [CrossRef]
15. Kumarage, S.P.; Dimantha De Silva, J. *Identification of Road Bottlenecks on Urban Road Networks Using Crowdsourced Traffic Data*; Elsevier Publications: Amsterdam, The Netherland, 2018.

16. Song, T.J.; Williams, B.M.; Rouphail, N.M. Data-driven approach for identifying spatiotemporally recurrent bottlenecks. *IET Intell. Transp. Syst.* **2018**, *12*, 756–764. [CrossRef]

17. Zhang, J.B.; Song, G.H.; Yu, L.; Guo, J.F.; Lu, H.Y. Identification and characteristics analysis of bottlenecks on urban expressways based on floating car data. *J. Cent. South Univ.* **2018**, *25*, 2014–2024. [CrossRef]

18. Yang, Y.; Li, M.; Yu, J.; He, F. Expressway bottleneck pattern identification using traffic big data—The case of ring roads in Beijing, China. *J. Intell. Transp. Syst.* **2019**, *24*, 54–67. [CrossRef]

19. Smith, B.L.; Williams, B.M.; Oswald, R.K. Comparison of parametric and nonparametric models for traffic flow forecasting. *Transp. Res. Part C Emerg. Technol.* **2002**, *10*, 303–321. [CrossRef]

20. Zambrano-Martinez, J.L.; Calafate, C.T.; Soler, D.; Cano, J.C.; Manzoni, P. Modeling and characterization of traffic flows in urban environments. *Sensors* **2018**, *18*, 2020. [CrossRef]

21. Ma, X.; Yu, H.; Wang, Y.; Wang, Y. Large-scale transportation network congestion evolution prediction using deep learning theory. *PLoS ONE* **2015**, *10*, e0119044. [CrossRef]

22. Wang, J.; Gu, Q.; Wu, J.; Liu, G.; Xiong, Z. Traffic speed prediction and congestion source exploration: A deep learning method. In Proceedings of the 2016 IEEE 16th International Conference on Data Mining (ICDM), Barcelona, Spain, 12–15 December 2016; pp. 499–508.

23. Hochreiter, S.; Schmidhuber, J. Long short-term memory. *Neural Comput.* **1997**, *9*, 1735–1780. [CrossRef]

24. Zhao, Z.; Chen, W.; Wu, X.; Chen, P.C.; Liu, J. LSTM network: A deep learning approach for short-term traffic forecast. *IET Intell. Transp. Syst.* **2017**, *11*, 68–75. [CrossRef]

25. Xiangxue, W.; Lunhui, X.; Kaixun, C. Data-driven short-term forecasting for urban road network traffic based on data processing and LSTM-RNN. *Arab. J. Sci. Eng.* **2019**, *44*, 3043–3060. [CrossRef]

26. Sun, W.; Wu, X.; Wang, Y.; Yu, G. A continuous-flow-intersection-lite design and traffic control for oversaturated bottleneck intersections. *Transp. Res. Part C Emerg. Technol.* **2015**, *56*, 18–33. [CrossRef]

27. Daganzo, C.F. Urban gridlock: Macroscopic modeling and mitigation approaches. *Transp. Res. Part B Methodol.* **2007**, *41*, 49–62. [CrossRef]

28. Daganzo, C.F.; Gayah, V.V.; Gonzales, E.J. Macroscopic relations of urban traffic variables: Bifurcations, multivaluedness and instability. *Transp. Res. Part B Methodol.* **2011**, *45*, 278–288. [CrossRef]

29. Saidallah, M.; El Fergougui, A.; Elalaoui, A.E. A comparative study of urban road traffic simulators. In *MATEC Web of Conferences*; EDP Sciences: Paris, France, 2016.

30. Lopez, P.A.; Behrisch, M.; Bieker-Walz, L.; Erdmann, J.; Flötteröd, Y.P.; Hilbrich, R.; Lücken, L.; Rummel, J.; Wagner, P.; Wießner, E. Microscopic Traffic Simulation using SUMO. In Proceedings of the 21st IEEE International Conference on Intelligent Transportation Systems, Maui, HI, USA, 4–7 November 2018.

31. Krajzewicz, D.; Erdmann, J.; Behrisch, M.; Bieker, L. Recent development and applications of SUMO-Simulation of Urban MObility. *Int. J. Adv. Syst. Meas.* **2012**, *5*, 128–138.

32. Yin, J.; Rao, W.; Yuan, M.; Zeng, J.; Zhao, K.; Zhang, C.; Li, J.; Zhao, Q. Experimental Study of Multivariate Time Series Forecasting Models. In Proceedings of the 28th ACM International Conference on Information and Knowledge Management, Beijing, China, 3–7 November 2019; pp. 2833–2839.

33. Brownlee, J. *Deep Learning for Time Series Forecasting: Predict the Future with MLPs, CNNs and LSTMs in Python*; Machine Learning Mastery: Vermont Victoria, Australia, 2018.

34. Chollet, F. Keras. 2015. Available online: https://keras.io (accessed on 1 October 2019).

35. Kingma, D.P.; Ba, J. Adam: A method for stochastic optimization. *arXiv* **2014**, arXiv:1412.6980.

36. Hyndman, R.J.; Koehler, A.B. Another look at measures of forecast accuracy. *Int. J. Forecast.* **2006**, *22*, 679–688. [CrossRef]

37. Hsueh, Y.L.; Chen, H.C. Map matching for low-sampling-rate GPS trajectories by exploring real-time moving directions. *Inf. Sci.* **2018**, *433*, 55–69. [CrossRef]

Article

Context-Aware Link Embedding with Reachability and Flow Centrality Analysis for Accurate Speed Prediction for Large-Scale Traffic Networks

Chanjae Lee [1] and Young Yoon [2,3,*]

1   Department of Artificial Intelligence, Big Data, Hongik University, 94 Wausan-ro, Sangsu-dong, Mapo-gu, Seoul 04068, Korea; arisel117@gmail.com
2   Department of Computer Engineering, Hongik University, 94 Wausan-ro, Sangsu-dong, Mapo-gu, Seoul 04068, Korea
3   Neouly Incorporated, 94 Wausan-ro, Sangsu-dong, Mapo-gu, Seoul 04068, Korea
*   Correspondence: young.yoon@hongik.ac.kr; Tel.:+82-2-320-1659

Received: 26 September 2020; Accepted: 28 October 2020; Published: 29 October 2020

Abstract: This paper presents a novel method for predicting the traffic speed of the links on large-scale traffic networks. We first analyze how traffic flows in and out of every link through the lowest cost reachable paths. We aggregate the traffic flow conditions of the links on every hop of the inbound and outbound reachable paths to represent the traffic flow dynamics. We compute a new measure called traffic flow centrality (i.e., the Z value) for every link to capture the inherently complex mechanism of the traffic links influencing each other in terms of traffic speed. We combine the features regarding the traffic flow centrality with the external conditions around the links, such as climate and time of day information. We model how these features change over time with recurrent neural networks and infer traffic speed at the subsequent time windows. Our feature representation of the traffic flow for every link remains invariant even when the traffic network changes. Furthermore, we can handle traffic networks with thousands of links. The experiments with the traffic networks in the Seoul metropolitan area in South Korea reveal that our unique ways of embedding the comprehensive spatio-temporal features of links outperform existing solutions.

Keywords: link embedding; traffic speed prediction; traffic flow centrality; reachability analysis; spatio-temporal data; artificial neural network; deep learning; context-awareness

---

1. Introduction

Accurate prediction of speed on traffic networks helps improve traffic management strategies and generate efficient routing plans. However, precisely estimating the traffic speed in advance has been a non-trivial task since various factors determine traffic flows in many ways.

Various approaches have been used for traffic speed prediction using statistical methods [1–7] and machine learning with neural networks with deep hidden layers [8–18]. However, the existing solutions are mostly limited to estimating traffic speed for either a single road link or a small-scale sub-network with only a handful of traffic links (e.g., a crossroad). In reality, a road system is a large-scale connected graph with the traffic links affecting each other's traffic flow over time in a more complicated fashion that cannot be explained easily with a simple statistical model. The works without a broader view of the traffic links may not adequately unravel the hidden, but critical speed prediction factors.

In this paper, we adapt the node embedding techniques introduced by Hamilton et al. [19]. We represent the relationship between links with a feature vector whose size is invariant even when certain changes are made to the traffic networks' structure, such as the addition or deletion of roads.

We analyze how the traffic flows in and out of a link through reachable multi-hop paths computed with the Floyd–Warshall algorithm [20]. How the links impact each other given the traffic flow analysis is quantified through a novel metric we refer to as Traffic Flow Centrality (TFC). We combine the data related to TFC with the external conditions around each link, such as climate and time information (e.g., time of day, the indication of holidays). We use a recurrent neural network algorithm to correlate between the composite feature's temporal transition and the traffic speed of each link. Our method captures the traffic flow dynamics through sub-networks around every link without embedding the entire adjacency matrix. Therefore, our method is much more space efficient, and it can handle large-scale traffic networks with thousands of links. We also do not face the problem of considering irrelevant information such as the hollow points around traffic networks when they are represented either with a sparse adjacency matrix or raster graphics [10,18]. Our solution assesses the impact of the remote links in addition to the adjacent neighbors on traffic flows. Hence, with the broader contextual view and the exclusion of unnecessary information, we expect to outperform the works that based their speed prediction only on the limited view of adjacent links.

This paper is structured as follows: In Section 2, we first review the related work. In Section 3, we introduce the method for inferring traffic speed given the temporal transitions of links' composite contextual features that are modeled with a novel link embedding technique. In Section 4, we benchmark the performance of our approach against existing works, and we conclude in Section 5.

2. Related Works

In this section, we put our work in the context of various related research works on traffic speed prediction. Recently, artificial neural networks with deep hidden layers have gained popularity, as they are effective at modeling the non-linear traffic speed dynamics. Some of the notable works have employed FNNs (Fuzzy Neural Networks) [8,9], DNNs (Deep Neural Networks) [11,21], RNNs (Recurrent Neural Networks) [13,14], DBNs (Deep Belief Networks) [12,15], and the IBCM-DL (Improved Bayesian Combination Model with Deep Learning) [17] models. These works have shown more accurate speed prediction than the approaches that are based on classic statistical methods such as ARIMA (Auto-Regressive Integrated Moving Average) models [1–3], SVR (Support Vector Regression) [4,5], and K-NN (K-Nearest Neighbor) [6,7].

When the models are obtained by learning the pattern on a single specific link [1–3,13–15,17], the distinct features of other links may not be adequately accounted for. Thus, modeling the traffic speed pattern per individual link was discussed in [22]. Nonetheless, the work by Kim et al. [22] still did not reflect the substructure of the traffic network around each link.

The works presented in [10,16,23,24] extracted spatial features from a visual representation of traffic networks. In particular, Zheng et al. [23] used a two-dimensional traffic flow that is embedded in Convolutional Neural Networks (CNN). They also used a Long Short-term Memory (LSTM) [25] algorithm to model long-term historical data. Similarly, Du et al. [16] represented the passenger flows among different traffic lines in a transportation network into multi-channel matrices with deep irregular convolutional networks. Guo et al. [10] analyzed the congestion propagation patterns based on the traffic observations recorded at fixed intervals in time and fixed locations in space. They represented the traffic observation in raster data. Then, all the raster data were fed into a 3D convolutional neural network to model the spatial information. They used a 3×3 convolutional kernel that includes the hollow points where no road lies and no traffic flows. The hollow points in the convolutional kernel may accidentally reflect stale traffic flows. Instead, Du et al. [16] used an irregular convolution kernel that refers to the traffic flow values from the adjacent traffic lines to fill in the values of the hollow points. These methods commonly exploit the CNN architecture that effectively models visual imagery [26–28]. Furthermore, a family of RNN algorithms such as GRU [29] and LSTM [25] was used so that repetitive temporal patterns can be discovered. However, these works did not capture the relation between the flow points on the two-dimensional spaces such as junctions, crossroads, and overpasses. Therefore, these models are susceptible to errors by correlating between irrelevant traffic flows. For instance, they may confuse overpasses and overlapping roads as crossroads. Furthermore, these works did not address the impact of the external conditions on the traffic flow. Learning the correlation between traffic flows and weather parameters was useful for flow prediction, as presented in [12,30]. However, they overlooked the impact of the substructure around traffic links on traffic flows.

More recently, modeling the traffic flow based on the graph representation of the traffic networks has emerged. The works presented in [18,31–33] combined GNNs (Graph Neural Networks) [34,35] and RNNs to capture the temporal flow transition patterns given adjacency matrices that explicitly reflect the complex interconnections. These methods do not have to unnecessarily deal with the information irrelevant to the traffic flow, such as the hollow points in the visual traffic networks that Du et al. [16] had to consider forcefully.

ST-TrafficNet [33] used Caltrans PeMS (Performance Measurement System) data from around 20 links between intersection points and predicted traffic speed with stacked LSTM using a spatially-aware multi-diffusion convolution block. This PeMS data were from 350 loop detectors at 5 min intervals from 1 January 2017 to 31 May 2017. This model reflects spatial influence through multi-diffusion convolution with forward, backward, and attentive channels.

TGC-LSTM was used by Cui et al. [18] to predict the traffic speed on four connected freeways in the Greater Seattle Area. They used publicly available traffic state data from 323 sensor stations over the entirety of 2015 at 5 min intervals. With their model, traffic speed was predicted with the RMSE (Root Mean Squared Error) as low as 2.1. However, the GNN architecture has to be restructured whenever the traffic networks undergo some changes. This is because TGC-LSTM uses the entire adjacency matrix as an input to the GNN instead of embedding the features of individual traffic links. Upon any change to the traffic networks, we have to re-train from scratch with the newly updated GNN architecture. Furthermore, since TGC-LSTM uses a very large adjacency matrix, both the time and space complexity of modeling the network structure becomes high. However, more importantly, the larger the adjacency matrix is, the more sparse it becomes. Therefore, TGC-LSTM still faces the problem of incorporating unnecessary data such as the hollow points captured in a regular convolution kernel, as discussed in [10]. The shortcomings of these GNN-based approaches motivated us to devise a new method for embedding the characteristics of the traffic network.

We adapt the node embedding techniques introduced by Hamilton et al. [19]. We represent the relationship between links on the traffic network with a feature vector whose size is invariant even when any part of the network structure changes. We analyze how the traffic routes through a link via reachable lowest cost multi-hop paths that are computed with the Floyd–Warshall algorithm [20]. We compute every link's relative impact on other links based on its inbound/outbound traffic flow patterns and its neighbors' collective conditions. We refer to the relative cross-link impact value as Traffic Flow Centrality (TFC). We combine the features related to TFC with the external conditions around each link, such as climate and time information. We use a recurrent neural network algorithm to learn how such a composite feature change over time determines the traffic speed of each link.

Our method does not involve the process of embedding the entire adjacency matrix. Therefore, our solution is more space efficient and can easily handle large-scale traffic networks with thousands of links. Furthermore, it avoids incorporating irrelevant information such as the hollow points that can be present in traffic networks when they are represented with a sparse adjacency matrix or raster graphics [10,18]. Our solution considers the conditions of the remote links beyond the adjacent neighbors. By ruling out irrelevant information and having the broader contextual view, we expect to outperform the works that base their speed prediction myopically on the conditions of the adjacent links.

The advantage of our work, named TFC-LSTM, is summarized in Table 1, which shows the comparison between existing related works we have discussed so far. The "Traffic Network Structure" refers to the usage of the abstract representation of interconnections between links. The "Surrounding Conditions" refer to the consideration of external situations around links such as climate and time information. The "Traffic Flow Reachability Analysis" refers to the process of analyzing the pattern of traffic flowing in and out of links through reachable paths. The "Centrality Analysis" refers to the usage of the link's relative influence on others. The "Chains of Neighbors" column indicates the consideration of remote neighbors besides the adjacent ones when capturing the substructure around a link.

Table 1. Comparison with other models. TFC, Traffic Flow Centrality.

Prediction Models	Raw Data Usage			Traffic Network Representation			
	Traffic Speed/Volume	Traffic Network Structure	Surrounding Conditions	Invariant Input Feature Vector Size	Traffic Flow Reachability Analysis	Centrality Analysis	Chains of Neighbors
TFC-LSTM	O	O	O	O	O	O	O
TGC-LSTM [18]	O	O	X	X	X	X	X
ST-3DNET [10]	O	O	X	X	X	X	X
DST-ICRL [16]	O	O	O	X	X	X	X
IBCM-DL [17]	O	X	X	X	X	X	X
DBN [15]	O	X	X	X	X	X	X
LSTM [13]	O	X	X	X	X	X	X
GRU [14]	O	X	X	X	X	X	X
DNN [12,22]	O	X	O	X	X	X	X
FNN [8,9]	O	O	X	X	X	O	X
K-NN [6,7]	O	O	X	X	O	X	X
SVR [4,5]	O	O	X	X	X	X	X
ARIMA [1–3]	O	X	X	X	X	X	X

"X" means no, and "O" means yes.

3. Methodology

We outline the overall procedure for predicting the link's speed on the traffic networks, as shown in Figure 1. Given raw data such as adjacency, distance, and speed in a data tensor, we compute reachability information such as hop count, distance, and cost of traffic flowing between a pair of source and destination links through a chain of neighboring links. Paired with other external features such as weather conditions, time of day, and day of the week, we expect that encoding the dynamics of the temporal traffic flows on the neighboring links would significantly improve the prediction of links' speed. How we generate the embedding of complex context-aware spatio-temporal features is explained in greater detail in Section 3.1. We model the correlation between the input feature and each link speed with a Long Short-Term Memory (LSTM) algorithm. We use 512 perceptrons in the hidden layer, ReLU for the activation function [36,37], and Adam for the optimizer [38,39].

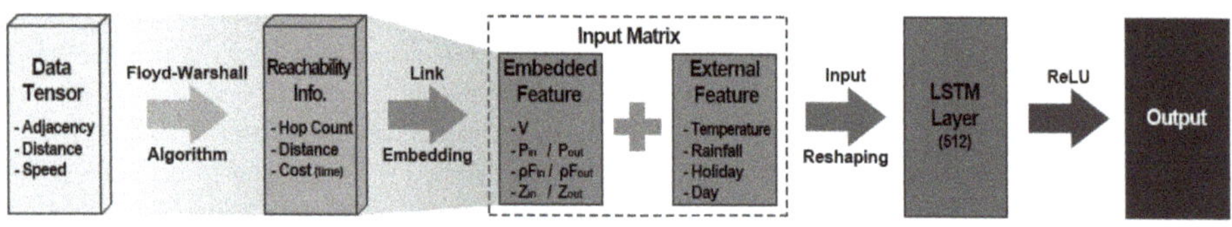

Figure 1. The overall procedure for traffic link embedding and training for traffic speed prediction.

3.1. Link Embedding

Feature representation of a link is done in the following three steps: First, we abstract the traffic system as a link adjacency matrix A. A traffic network illustrated in Figure 2A is abstracted further as a directed graph structure, as shown in Figure 2B, with every directional link being uniquely identified. Figure 3A shows the link adjacency matrix for the sample traffic system in Figure 2A. $A_{l_i l_j}$ denotes whether traffic can flow from link l_i to link l_j. For instance, according to the example in Figure 2A,

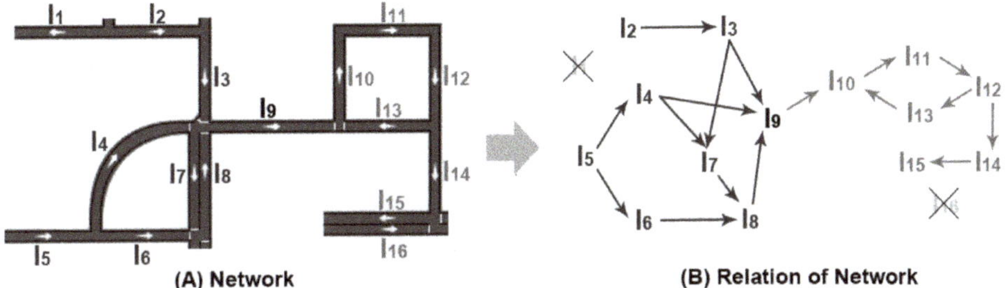

Figure 2. Relation and bound of the network.

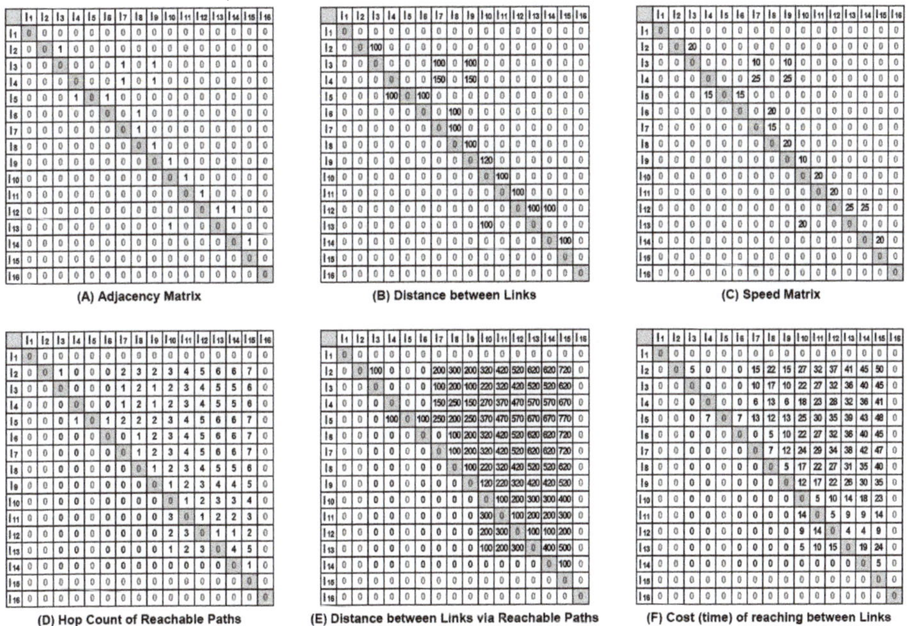

Figure 3. Matrices representing adjacency, distance, and information about reachable paths between links.

$A_{l_2 l_3} = 1$ since traffic can flow from l_2 to l_3. Otherwise, $A_{l_2 l_3}$ equals 0. We exclude the links through which no traffic can reach any other parts of the traffic system. For instance, $l1$ and $l16$ are disconnected from the rest of the traffic system. Figure 3B illustrates a matrix of the distance between every pair of adjacent links. Given $A_{l_i l_j}$, Figure 3C shows the speed of the traffic flowing from l_i to adjacent link l_j.

Second, we conduct a traffic flow reachability analysis. For every link l, we extract all inbound multi-hop paths through which traffic traverses towards l. Oppositely, we compute all outbound multi-hop paths through which the traffic originated from l diffuses. Given $A_{l_i l_j}$, we run the Floyd–Warshall algorithm [20] to obtain the minimum time it takes to travel from every source link l_i to all other destination links l_j, as shown in Figure 3F. We generate a matrix of hop counts and minimum distances from every source link l_i to all other destination link l_j, as shown in Figure 3D,E, respectively. For instance, the traffic on l_5 (at Hop 3) takes l_6 (at Hop 2) to reach l_9 at time t_2 in Figure 4B as opposed to taking l_4 (at Hop 1) at time t_1 in Figure 4A due to the heavy congestion on l_4. After we obtain the paths, we discard the links through which traffic cannot reach the source link before the beginning of the next time window. For example, if the traffic on l_9 is too slow to reach the remote link

l_{15} by the end of the current time window, then we discard l_{15} from the reachable outbound path from l_9, as shown in Figure 5.

$$P_{in} = \sum_{i \in R}^{R} C_{in} \quad , \quad P_{out} = \sum_{i \in R}^{R} C_{out} \tag{1}$$

$$\rho F_{in} = \sum_{i \in R}^{R} \frac{v_{in_i}}{f_{in_i} d_{in_i}} \quad , \quad \rho F_{out} = \sum_{i \in R}^{R} \frac{v_{out_i}}{f_{out_i} d_{out_i}} \tag{2}$$

$$L_{x_{in}} = \frac{\rho F_{in}}{\sum_{i \in R}^{R} \rho F_{in}} \quad , \quad L_{x_{out}} = \frac{\rho F_{out}}{\sum_{i \in R}^{R} \rho F_{out}} \tag{3}$$

$$Z_{in} = \ln\left(\sum_{n \in N_{in}}^{N_{in}} \frac{L_{n_{in}}}{e^{hop}} \right) , \quad Z_{out} = \ln\left(\sum_{n \in N_{out}}^{N_{out}} \frac{L_{n_{out}}}{e^{hop}} \right) \tag{4}$$

$$Z_{in} = \frac{Z_{n_{in}}}{\sum_{n \in N}^{N} Z_{n_{in}}} , \quad Z_{out} = \frac{Z_{n_{out}}}{\sum_{n \in N}^{N} Z_{n_{out}}} \tag{5}$$

$$|Z_{in} - Z'_{in}| < \epsilon \, , \, |Z_{out} - Z'_{out}| < \epsilon \tag{6}$$

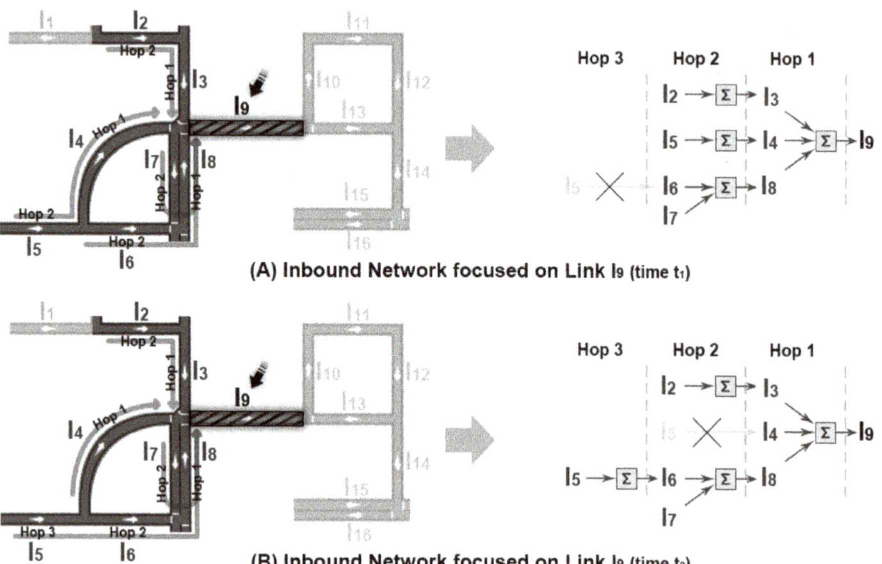

Figure 4. Extraction of inbound reachable paths and the computation of the inbound Z value for I_9.

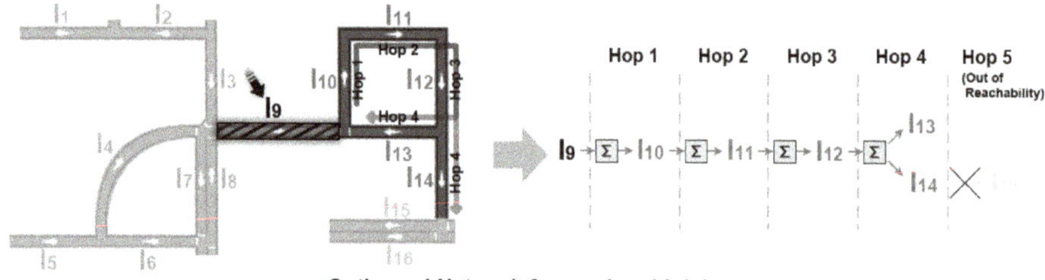

Figure 5. Extraction of outbound reachable paths and the computation of the outbound Z value for I_9.

In the final step, we compute the Z value for every link, which we refer to as the traffic flow centrality. Given the matrix of minimum time to travel from source link l_i to destination link l_j (Figure 6A), we count the number of inbound and outbound reachable paths (R), $(P_{in}$ and $P_{out})$for every link using Equation (1), as illustrated in Figure 6B. We also compute the speed and distance between every pair of adjacent links on the reachable paths. Equation (2) defines (ρF) as the weighted sum of the average traffic speed (v) on every link i on the multi-hop reachable paths. The weight of each intermediate link i is the inverse of the product between the fanout f and the distance d to i, where f specifically represents the number of alternate paths on a junction. The weight represents the impact on a given link the current traffic is either destined to or stemmed from. We expect the impact to be sensitive to the f and d values. For instance, we can capture the circumstance where the traffic on links with higher f and d values are less likely to move towards a target link l than the traffic on a link with lower f and d values. This is because traffic on the link with higher f and d values is more likely to veer away by taking different turns on the junctions or halt the transition at any point on the path. As an example, computing the inbound and the outbound ρF values for the link l_9 is illustrated in Figure 6C,D. The aggregation steps above are to account for the traffic flow dynamics around every link.

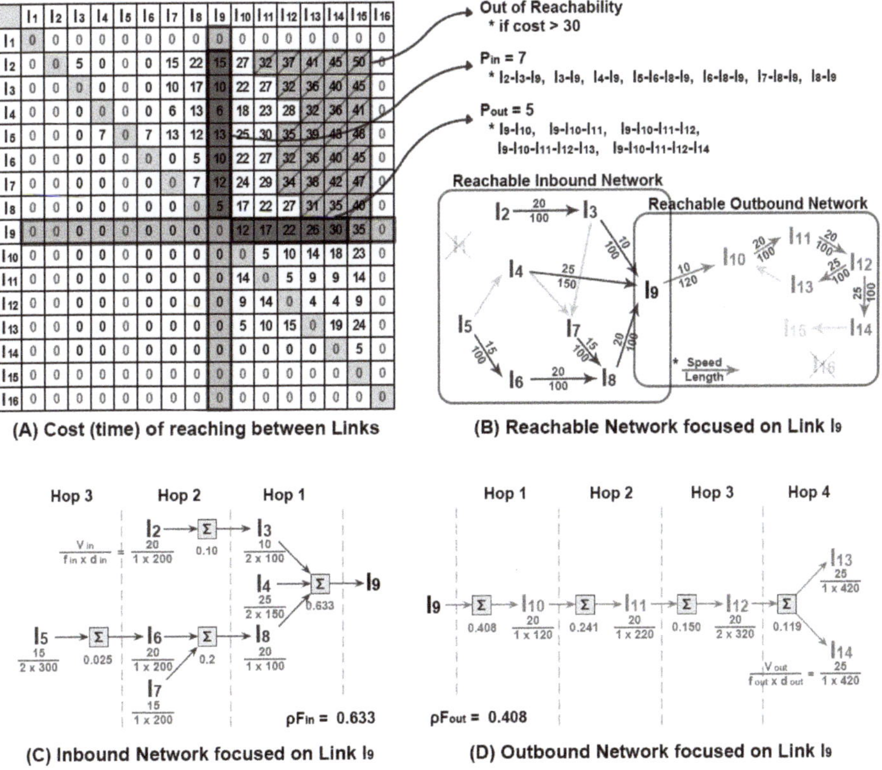

Figure 6. An example of aggregating the traffic conditions of the neighboring links on reachable paths.

Note that the links adjacent to each other are also inter-dependent on each other with regards to computing their Z values. Due to the inter-dependence, we have to compute Equations (4) and (5) iteratively until the condition on Equation (6) holds. We obtain a converged Z value when Equation (6) is satisfied. We re-scale the ρF values to L_x through normalization as specified in Equation (3). Suppose we have a set of adjacent inbound and outbound neighbors N_{in} and N_{out}, respectively, for a given link. Then, we take the sums of L_x/e^{hop} of every neighbor in N_{in} and N_{out}. The division by e^{hop} reflects that the impact of a link's L_x value on its immediate neighbor is inversely proportional to the hop distance between them. We take the natural logarithmic function on the sums as shown in Equation (4) and normalize the value as defined in Equation (5). Whenever Equation (4) repeats, L_n is substituted by Z' obtained in the previous iteration. The ϵ value in Equation (6) is for determining the converged Z. We empirically set the ϵ value that leads to the most accurate link speed prediction. The iterative computation of the Z value for every link in a traffic network is illustrated in Figure 7.

The traffic flow centrality is to capture the inter-link relationship, and we expect it to be one of the key factors for accurately predicting traffic speed. Intuitively, it is highly probable that a link would experience traffic swarming in and causing congestion, especially when a large portion of its immediate neighbors also experience high inbound traffic through the reachable paths. Furthermore, a link with several surrounding links that spread out traffic quickly is likely to disseminate its traffic more easily.

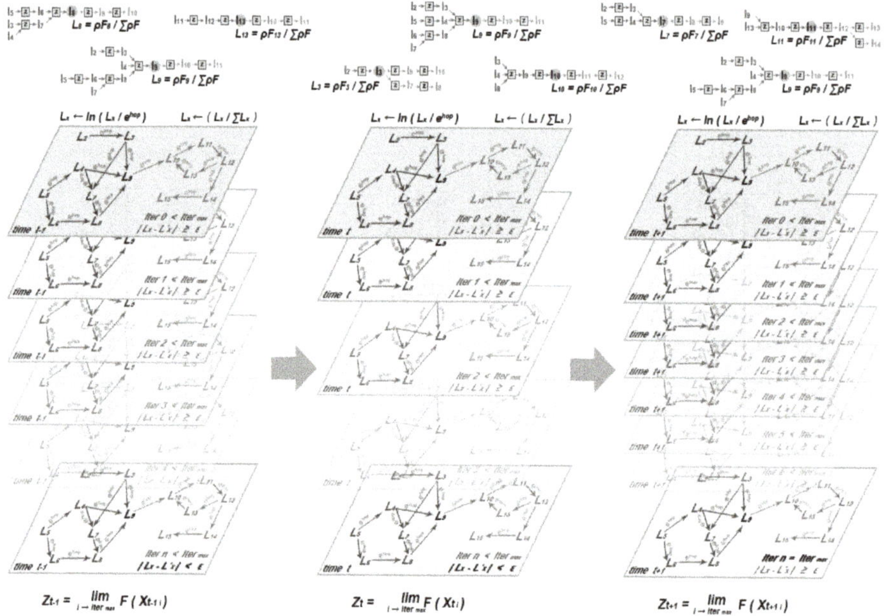

Figure 7. Iterative computation of traffic flow centrality for every link in the traffic network.

We embed the 7 features, $(V, P_{in}, P_{out}, \rho F_{in}, \rho F_{out}, Z_{in}, Z_{out})$, into a vector for every link, where V is the traffic speed. We make this link embedding more context-aware by simply concatenating an additional vector of external conditions surrounding the link. The external conditions include temperature, precipitation, time of day, day of the week, and an indication of whether a given day is a public holiday. The steps for generating the final input matrix we feed into a neural network for speed prediction is illustrated in Figure 8.

No matter how many adjacent neighbors a link has, the length of the input vector remains invariant, thus making our solution resilient to changes such as the addition or deletion of neighboring links. We do not take the entire adjacency matrix as an input to the prediction engine based on recurrent neural networks. Therefore, our approach becomes space-efficient. Our input vectors contain only the essential information instead of the initial raw adjacency matrix that is sparse. With every link expressed with highly distinct features, we expect it to yield better prediction results. Note that the input vectors are computed at every time window. In the following section, we introduce the method for modeling the transition of the features over time.

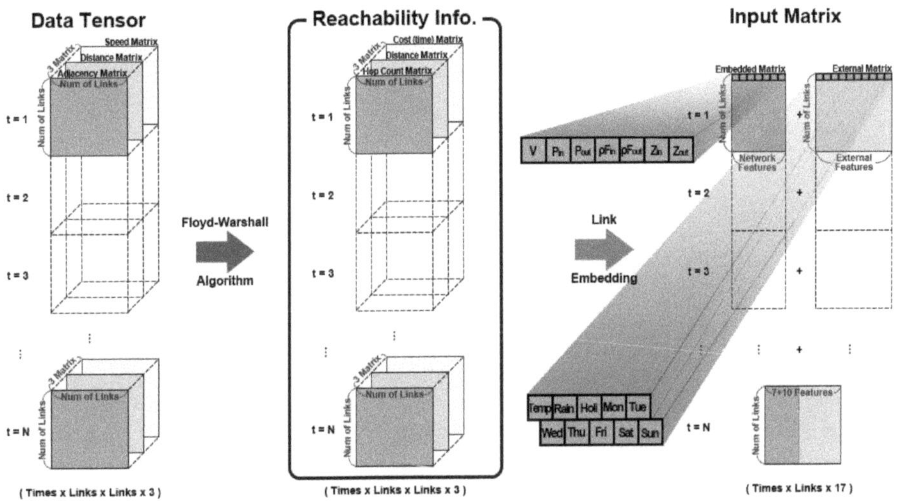

Figure 8. Generation of the input tensor reflecting the context-aware traffic flow centrality of every link.

3.2. Modeling the Temporal Patterns with Recurrent Neural Networks

Given the input matrix generated at a time window, we predict the traffic speed of every link in the next time window using recurrent neural networks, such as GRU and LSTM. With the hidden layers, we can model a complex non-linear relationship between the input and the output values. Specifically, the example of feeding in the input features of every link and predicting its speed through an LSTM network is illustrated in Figure 9. The states summarized in the previous time window are fed into the block of hidden layers at the subsequent time window. Besides the previous states, we feed in the updated input feature vectors of every link. By training this network, we can model the spatial information's temporal transition, such as the transition of the traffic system's structure, traffic flow dynamics, and external conditions. We also make the correlation between the traffic speed of each link and the temporal transition of the most comprehensive set of features to date.

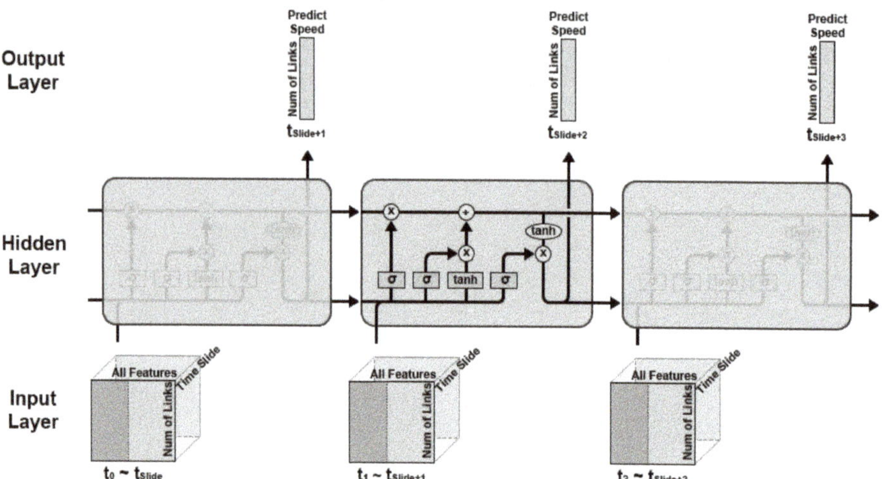

Figure 9. Using LSTM for modeling the correlation between the temporal patterns and the speed prediction at every time window.

4. Evaluation

In this section, we evaluate our approach using the data from TOPIS (Seoul Traffic Operation and Information service) https://topis.seoul.go.kr/, which contain hourly average speed information for the 4670 major traffic links in the Seoul metropolitan area. We obtained data at one hour intervals for 8760 h worth of data from 00:00 on 1 January 2018 to 23:00 on 31 December 2018. Temperature and precipitation information of each link was obtained from the nearest weather station. We retrieved the climate readings from KMA's https://data.kma.go.kr/ AWS(Automatic Weather System). We took 60% of the data as a training set, 20% as a validation set, and the rest as the test set. The attributes of the dataset used in this paper are listed in Table 2 with basic statistics, units, measurement intervals, and data types.

Table 2. Data attributes.

Data	Speed	Temperature	Precipitation	Day	Holiday
Mean	28.17	12.88	0.14	-	-
Median	25.62	13.7	0	-	-
Max	110.00	40.5	76	1	1
Min	0.74	−27.1	0	0	0
Unit	km/h	°C	mm	-	-
Time Interval	1 h	1 h	1 h	1 day	1 day
Type	float	float	float	binary	binary

We conducted our experiments on NVIDIA DGX-1 with an 80-Core (160 threads) CPU, 8 Tesla V100 GPUs with 32 GB of exclusive memory and 512 GB of RAM. NVIDIA DGX-1 was operated with the Ubuntu 16.04.5 LTS server, and the machine learning tasks were executed through the Docker containers. The machine learning algorithms were implemented with Python (v3.6.9), Tensorflow (v1.15.0), and Keras (v2.3.1) libraries. We used ReLU for the activation function [36,37] and Adam for the optimization function [38,39]. The learning rate was empirically set to 0.001. The early stoppage was configured by setting the patience value to 200 through Keras. To measure the prediction performance, we employed the metrics as follows: (1) MAE (Mean Absolute Error) (Equations (7)); (2) RMSE (Root Mean Squared Error) (Equation (8)); and (3) MAPE (Mean Absolute Percentage Error) (Equation (9)). y_t and $\hat{y}_t$ are the predicted speed and the actual speed, respectively. n is the number of test cases. The ϵ value for checking the convergence of the Z value, as defined in Equation (6), was empirically set as shown in Table 3. With $\epsilon = 0.005$, we were able to achieve the lowest MAPE.

$$MAE = \frac{1}{n} \sum_{t=1}^{n} |y_t - \hat{y}_t| \tag{7}$$

$$RMSE = \sqrt{\frac{1}{n} \sum_{t=1}^{n} (y_t - \hat{y}_t)^2} \tag{8}$$

$$MAPE = \frac{1}{n} \sum_{t=1}^{n} \left| \frac{y_t - \hat{y}_t}{y_t} \right| * 100 \quad (\%) \tag{9}$$

Table 3. Prediction performance according to the convergence condition (ϵ) setting.

Epsilon	MAE	RMSE	MAPE
0.1	2.60	4.40	10.84
0.05	2.69	4.48	11.16
0.01	2.63	4.41	11.07
0.005	2.50	4.28	10.39
0.001	2.69	4.43	11.15
0.0005	2.59	4.36	10.79
0.0001	2.54	4.32	10.51

4.1. Measurement of Prediction Performance

We measured the prediction performance between various approaches denoted as DNN, TFC-DNN, GRU, TFC-GRU, LSTM, and TFC-LSTM. The prefix TFC- stands for Traffic Flow Centrality, and it represents the following seven novel features that we introduced in Section 3, i.e., V, P_{in}, P_{out}, ρF_{in}, ρF_{out}, Z_{in}, and Z_{out}. We evaluated the effectiveness of the information about external conditions such as climate and date information separately. DNN, GRU, and LSTM are the artificial neural network architectures we employed for machine learning. Table 4 contains the prediction performance of each approach. For each model, we picked the empirically best hyper-parameter settings, such as the number of hidden layers, perceptrons per layer, and the number of time windows for the recurrent neural networks. For double hidden layers, we had 64 and eight perceptrons for the first and the second layer, respectively. For a single hidden layer, we used 512 perceptrons. We cited the representative existing works that fall under the prediction model categories that do not use our techniques. We did not compare the methods that could not deal with the traffic networks that scale to thousands of links. MSE values during the training and validation stage are plotted on the graphs in Figure 10. MSEs converged at epoch = 400.

The consideration of every link's external conditions was effective only when used by LSTM and TFC-LSTM. On the other hand, we observed that using the features related to traffic flow centrality consistently led to improvement over the baseline approaches. However, when both the external conditions and the features related to traffic flow centrality were used with LSTM, i.e., TFC-LSTM, we achieved the lowest MAPE of 10.39.

Table 4. Comparison of prediction performance.

Prediction Models	Usage of Surrounding Condition Information	Hidden Layer Structure (# of Perceptrons per Layer)	# of Time Windows	MAE	RMSE	MAPE
DNN [12,22]	No	Double layers (64-8)	-	4.05	6.25	15.58
	Yes	Double layers (64-8)	-	4.36	6.42	17.36
TFC-DNN	No	Double layers (64-8)	-	3.84	5.98	15.27
	Yes	Double layers (64-8)	-	3.96	6.01	17.02
GRU [14]	No	Single layer (512)	18	3.70	5.51	14.42
	Yes	Single Layer (512)	18	7.52	10.20	31.12
TFC-GRU	No	Single layer (512)	18	2.85	4.60	11.42
	Yes	Single Layer (512)	18	6.20	8.12	22.57
LSTM [13]	No	Single layer (512)	18	2.88	4.91	11.89
	Yes	Single Layer (512)	18	2.64	4.48	10.87
TFC-LSTM	No	Single layer (512)	18	2.63	4.47	11.38
	Yes	Single layer (512)	18	2.50	4.28	10.39

The DNN-based approaches without the information about the external conditions performed poorly compared to the recurrent neural network models because it does not model the temporal transitions of the link features.

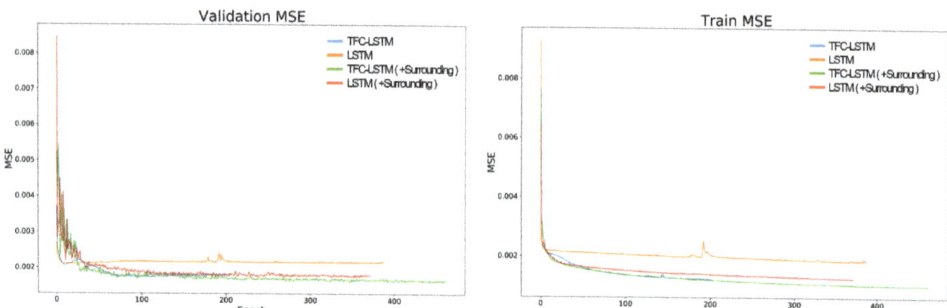

Figure 10. MSEs during training and validation.

4.2. The Effect of Reachable Path Length Cutoff

We noticed that a link in the Seoul traffic system could be reached from any other links within an hour regardless of the distance, even during rush hours with heavy traffic, according to the Floyd–Warshall algorithm we used for retrieving the lowest cost inbound and outbound paths between any OD pairs. It turns out that the Floyd–Warshall algorithm retrieved several unrealistic reachable paths between any OD pairs by ignoring specific restrictions on some of the traffic links. For example, the algorithm generated some routes that included wild U-turns that are not allowed on some roads. As a result, path lengths measured as the number of hops tended to be excessively long. Thus, we were unnecessarily taking into account the state of the links for which it is practically impossible to influence the state of the remote links.

One possible solution to this problem is to have a shorter time window than an hour. However, TOPIS only makes the hourly data available to the public. Therefore, we revised the algorithm to limit the path length instead. When overall traffic flows at the lowest average speed, we limited the reachable inbound and outbound path lengths to 15 hops. On the other hand, for the period when overall traffic flows at the highest average speed, we relaxed the length limit to 45 hops. Furthermore, the link's low speed may be attributed to the congestion on the neighboring links in the vicinity. Thus, reachability from other links to the low-speed link is also limited. Therefore, it is sufficient to consider only the links in proximity. Compared to the period of lowest average traffic speed ("Lowest Speed

Period" in Table 5), the reachability of the traffic from one link to the others is higher during the highest average traffic speed period. Thus, for such a period ("Highest Speed Period" in Table 5), we considered the states of the other links within a wider range. Our simple revision of the algorithm is empirically proven to be effective, as shown in Table 5. It shows the prediction performance for different path length cutoff settings. We observed that our approach performed most effectively with an MAPE of 10.39 when the range of neighboring links to consider was proportional to the traffic's average speed.

Table 5. The effect of reachability path length limit on speed prediction.

Lowest Speed Period	Highest Speed Period	MAE	RMSE	MAPE
15 hops	15 hops	2.61	4.38	10.83
30 hops	30 hops	2.51	4.30	10.48
45 hops	45 hops	2.61	4.40	10.96
15 hops	45 hops	2.50	4.28	10.39
45 hops	15 hops	2.63	4.44	10.98

4.3. The Discussion on Scalability

One of the merits of our approach is the capability to predict the traffic speeds even for a very large-scale traffic network such as the road system of Seoul. The larger the traffic network is, the higher the opportunity is to predict speed accurately. This is because we can consider more conditions around the links that are often overlooked by the existing works. The works that only consider the conditions of adjacent neighbors [18] exhibited a decline in prediction accuracy compared to what was originally reported and performed worse than our approach. This is because their approaches were unableto capture the farther away links' highly probable influence.

Naively feeding in the raw adjacency matrix was the most space inefficient approach [40,41]. We could not even compare the prediction performance with such approaches, as they quickly encountered out-of-memory errors when dealing with Seoul's large-scale road system. Our approach is agnostic of the scale of the network and even the structure change. Regardless of the scale and any changes, we ran the aggregation functions to compute the fixed-length input feature vector concerning the traffic flow centrality. Therefore, we did not need to restructure the neural network architecture upon changes to the traffic system (i.e., addition/deletion of links, change of adjacent links).

However, we dealt with a fragment of the entire national traffic system in South Korea. The traffic networks in Seoul are connected to the systems in other districts such as Gyeonggi Province and Incheon metropolitan area. Due to the fragmented view, we accidentally identified the links that bridge between different traffic networks as dead-ends, as shown in Figure 11. We could not accurately reflect the traffic flow centrality for these dead-end links. The Z_{out} value was zero for these dead-end links because there is no way out. The Z_{in} value is not credible as the inbound traffic from other regions was not accounted for. The inaccurate Z values on the dead-end may negatively impact the neighboring links' Z values.

For this issue, we applied the average Z value of the whole system as the Z value of the dead-end links, which still cannot be viewed as an ideal solution. This motivated us to venture into applying our approach to predicting speed prediction for all the links in the entire national traffic system. By expanding the view of the traffic network, we expect the accuracy to improve further. This is planned to be done soon after we are given integrated traffic information from all Korean regions.

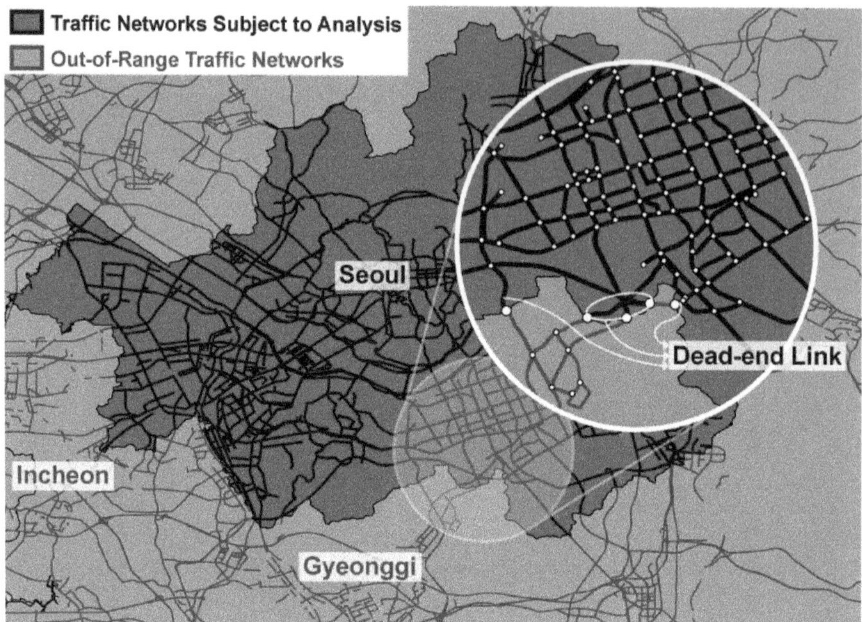

Figure 11. Bridge links between traffic networks subject to analysis and out-of-range traffic network accidentally being recognized as dead-end links.

5. Conclusions

We presented a novel method for describing the dynamic circumstances around any given traffic links. Specifically, by using a new measure called traffic flow centrality, we were able to concisely express the dynamics of the traffic flow in and out of any given link. Through this measure, we can reflect on the inherently complex ways in the interconnected traffic links affecting each other. Combined with the information about the external conditions surrounding the links, e.g., climate and time of day, the new features and their temporal patterns are used for predicting traffic speed. According to the information available from TOPIS (Seoul Traffic Operation and Information service), training with the LSTM algorithm given our comprehensive spatio-temporal features yielded the lowest prediction error with an MAPE of 10.39. Our experiment also shows that our solution is easily applicable to large-scale traffic systems. We could predict the traffic speed on the road network with thousands of links, unlike the existing works without any efficient feature embedding approaches.

As future work, we plan to apply our solution to the nation-wide traffic system.

Author Contributions: Conceptualization, Y.Y.; methodology, C.L. and Y.Y.; software, C.L.; validation, C.L. and Y.Y.; formal analysis, C.L. and Y.Y.; investigation, C.L.; resources, C.L.; data curation, C.L.; writing, original draft preparation, C.L. and Y.Y.; writing, review and editing, C.L. and Y.Y.; visualization, C.L.; supervision, Y.Y.; project administration, Y.Y.; funding acquisition, Y.Y. All authors read and agreed to the published version of the manuscript.

Funding: This research was supported by the Korean Agency for Infrastructure Technology Advancement (KAIA) grant funded by the Ministry of Land, Infrastructure, and Transport (Grant 20TLRP-B148970-03), by the Basic Science Research Programs through the National Research Foundation of Korea (NRF) funded by the Korea government (MSIT) (2020R1F1A104826411), by the Ministry of Health & Welfare, Republic of Korea (Grant Number HI19C0542020020), by the Ministry of Trade, Industry & Energy (MOTIE) and the Korea Institute for Advancement of Technology (KIAT), under Grants P0004602 and P0014268 Smart HVAC demonstration support, and by the 2020 Hongik University Research Fund.

Conflicts of Interest: The authors declare no conflict of interest. The funders had no role in the design of the study; in the collection, analyses, or interpretation of data; in the writing of the manuscript; nor in the decision to publish the results.

Abbreviations

The following abbreviations are used in this manuscript:

OD   Origin and Destination
TFC   Traffic Flow Centrality (Z value)

References

1. Williams, B.M.; Hoel, L.A. Modeling and Forecasting Vehicular Traffic Flow as a Seasonal ARIMA Process: Theoretical Basis and Empirical Results. *J. Transp. Eng.* **2003**, *129*, 664–672. [CrossRef]
2. Kumar, S.V.; Vanajakshi, L. Short-Term Traffic Flow Prediction using Seasonal ARIMA Model with Limited Input Data. *Eur. Transp. Res. Rev.* **2015**, *7*, 21. [CrossRef]
3. Chen, C.; Hu, J.; Meng, Q.; Zhang, Y. Short-Time Traffic Flow Prediction with ARIMA-GARCH Model. In Proceedings of the 2011 IEEE Intelligent Vehicles Symposium (IV), Baden-Baden, Germany, 5–9 June 2011; pp. 607–612.
4. Jeong, Y.S.; Byon, Y.J.; Castro-Neto, M.M.; Easa, S.M. Supervised Weighting-Online Learning Algorithm for Short-Term Traffic Flow Prediction. *IEEE Trans. Intell. Transp. Syst.* **2013**, *14*, 1700–1707. [CrossRef]
5. Wu, C.H.; Ho, J.M.; Lee, D.T. Travel-Time Prediction with Support Vector Regression. *IEEE Trans. Intell. Transp. Syst.* **2004**, *5*, 276–281. [CrossRef]
6. Cai, P.; Wang, Y.; Lu, G.; Chen, P.; Ding, C.; Sun, J. A Spatiotemporal Correlative K-Nearest Neighbor Model for Short-Term Traffic Multistep Forecasting. *Transp. Res. Part C Emerg. Technol.* **2016**, *62*, 21–34. [CrossRef]
7. Kim, T.; Kim, H.; Lovell, D.J. Traffic Flow Forecasting: Overcoming Memoryless Property in Nearest Neighbor Non-Parametric Regression. In Proceedings of the 2005 IEEE Intelligent Transportation Systems, 2005, Vienna, Austria, 16 September 2005; pp. 965–969.
8. Yin, H.; Wong, S.; Xu, J.; Wong, C. Urban Traffic Flow Prediction using a Fuzzy-Neural Approach. *Transp. Res. Part C Emerg. Technol.* **2002**, *10*, 85–98. [CrossRef]
9. Quek, C.; Pasquier, M.; Lim, B.B.S. POP-TRAFFIC: A Novel Fuzzy Neural Approach to Road Traffic Analysis and Prediction. *IEEE Trans. Intell. Transp. Syst.* **2006**, *7*, 133–146. [CrossRef]
10. Guo, S.; Lin, Y.; Li, S.; Chen, Z.; Wan, H. Deep Spatial-Temporal 3D Convolutional Neural Networks for Traffic Data Forecasting. *IEEE Trans. Intell. Transp. Syst.* **2019**, *20*, 3913–3926. [CrossRef]
11. Lv, Y.; Duan, Y.; Kang, W.; Li, Z.; Wang, F.Y. Traffic Flow Prediction with Big Data: A Deep Learning Approach. *IEEE Trans. Intell. Transp. Syst.* **2014**, *16*, 865–873. [CrossRef]
12. Koesdwiady, A.; Soua, R.; Karray, F. Improving Traffic Flow Prediction with Weather Information in Connected Cars: A Deep Learning Approach. *IEEE Trans. Veh. Technol.* **2016**, *65*, 9508–9517. [CrossRef]
13. Tian, Y.; Pan, L. Predicting Short-Term Traffic Flow by Long Short-Term Memory Recurrent Neural Network. In Proceedings of the 2015 IEEE International Conference on Smart City/SocialCom/SustainCom (SmartCity), Chengdu, China, 19–21 December 2015; pp. 153–158.
14. Fu, R.; Zhang, Z.; Li, L. Using LSTM and GRU Neural Network Methods for Traffic Flow Prediction. In Proceedings of the 2016 31st Youth Academic Annual Conference of Chinese Association of Automation (YAC), Wuhan, China, 11–13 November 2016; pp. 324–328.
15. Huang, W.; Song, G.; Hong, H.; Xie, K. Deep Architecture for Traffic Flow Prediction: Deep Belief Networks with Multitask Learning. *IEEE Trans. Intell. Transp. Syst.* **2014**, *15*, 2191–2201. [CrossRef]
16. Du, B.; Peng, H.; Wang, S.; Bhuiyan, M.Z.A.; Wang, L.; Gong, Q.; Liu, L.; Li, J. Deep Irregular Convolutional Residual LSTM for Urban Traffic Passenger Flows Prediction. *IEEE Trans. Intell. Transp. Syst.* **2019**, *21*, 972–985. [CrossRef]
17. Gu, Y.; Lu, W.; Xu, X.; Qin, L.; Shao, Z.; Zhang, H. An Improved Bayesian Combination Model for Short-Term Traffic Prediction with Deep Learning. *IEEE Trans. Intell. Transp. Syst.* **2019**, *21*, 1332–1342. [CrossRef]
18. Cui, Z.; Henrickson, K.; Ke, R.; Wang, Y. Traffic Graph Convolutional Recurrent Neural Network: A Deep Learning Framework for Network-Scale Traffic Learning and Forecasting. *IEEE Trans. Intell. Transp. Syst.* **2019**. [CrossRef]
19. Hamilton, W.L.; Ying, R.; Leskovec, J. Representation Learning on Graphs: Methods and Applications. *arXiv* **2017**, arXiv:1709.05584.
20. Floyd, R.W. Algorithm 97: Shortest Path. *Commun. ACM* **1962**, *5*, 345. [CrossRef]

21. Zhang, J.; Zheng, Y.; Qi, D.; Li, R.; Yi, X. DNN-based Prediction Model for Spatio-temporal Data. In Proceedings of the 24th ACM SIGSPATIAL International Conference on Advances in Geographic Information Systems, Burlingame, CA, USA, 31 October–3 November 2016; pp. 1–4.

22. Kim, D.H.; Hwang, K.Y.; Yoon, Y. Prediction of Traffic Congestion in Seoul by Deep Neural Network. *J. Korea Inst. Intell. Transp. Syst.* **2019**, *18*, 44–57. [CrossRef]

23. Zheng, Z.; Yang, Y.; Liu, J.; Dai, H.N.; Zhang, Y. Deep and Embedded Learning Approach for Traffic Flow Prediction in Urban Informatics. *IEEE Trans. Intell. Transp. Syst.* **2019**, *20*, 3927–3939. [CrossRef]

24. Sun, S.; Wu, H.; Xiang, L. City-wide traffic flow forecasting using a deep convolutional neural network. *Sensors* **2020**, *20*, 421. [CrossRef]

25. Hochreiter, S.; Schmidhuber, J. Long Short-Term Memory. *Neural Comput.* **1997**, *9*, 1735–1780. [CrossRef]

26. Krizhevsky, A.; Sutskever, I.; Hinton, G.E. Imagenet Classification with Deep Convolutional Neural Networks. In Proceedings of the Advances in Neural Information Processing Systems, Lake Tahoe, NV, USA, 3–6 December 2012; pp. 1097–1105.

27. Karpathy, A.; Toderici, G.; Shetty, S.; Leung, T.; Sukthankar, R.; Li, F.-F. Large-Scale Video Classification with Convolutional Neural Networks. In Proceedings of the IEEE conference on Computer Vision and Pattern Recognition, Columbus, OH, USA, 23–28 June 2014; pp. 1725–1732.

28. Ji, S.; Xu, W.; Yang, M.; Yu, K. 3D Convolutional Neural Networks for Human Action Recognition. *IEEE Trans. Pattern Anal. Mach. Intell.* **2012**, *35*, 221–231. [CrossRef]

29. Chung, J.; Gulcehre, C.; Cho, K.; Bengio, Y. Empirical Evaluation of Gated Recurrent Neural Networks on Sequence Modeling. *arXiv* **2014**, arXiv:1412.3555.

30. Jia, Y.; Wu, J.; Xu, M. Traffic Flow Prediction with Rainfall Impact using a Deep Learning Method. *J. Adv. Transp.* **2017**, *2017*. [CrossRef]

31. Yu, B.; Lee, Y.; Sohn, K. Forecasting Road Traffic Speeds by Considering Area-Wide Spatio-temporal Dependencies based on a Graph Convolutional Neural Network (GCN). *Transp. Res. Part C Emerg. Technol.* **2020**, *114*, 189–204. [CrossRef]

32. Ge, L.; Li, S.; Wang, Y.; Chang, F.; Wu, K. Global Spatial-Temporal Graph Convolutional Network for Urban Traffic Speed Prediction. *Appl. Sci.* **2020**, *10*, 1509. [CrossRef]

33. Lu, H.; Huang, D.; Song, Y.; Jiang, D.; Zhou, T.; Qin, J. St-trafficnet: A spatial-temporal deep learning network for traffic forecasting. *Electronics* **2020**, *9*, 1474. [CrossRef]

34. Scarselli, F.; Gori, M.; Tsoi, A.C.; Hagenbuchner, M.; Monfardini, G. The Graph Neural Network Model. *IEEE Trans. Neural Netw.* **2008**, *20*, 61–80. [CrossRef]

35. Kipf, T.N.; Welling, M. Semi-Supervised Classification with Graph Convolutional Networks. *arXiv* **2016**, arXiv:1609.02907.

36. Nair, V.; Hinton, G.E. Rectified Linear Units Improve Restricted Boltzmann Machines; In Proceedings of the 27th International Conference on Machine Learning (ICML 2010), Haifa, Israel, 21–24 June 2010.

37. Glorot, X.; Bordes, A.; Bengio, Y. Deep Sparse Rectifier Neural Networks. In Proceedings of the Fourteenth International Conference on Artificial Intelligence and Statistics, Fort Lauderdale, FL, USA, 11–13 April 2011; pp. 315–323.

38. Kingma, D.P.; Ba, J. Adam: A method for Stochastic Optimization. *arXiv* **2014**, arXiv:1412.6980.

39. Ruder, S. An Overview of Gradient Descent Optimization Algorithms. *arXiv* **2016**, arXiv:1609.04747.

40. Yu, B.; Yin, H.; Zhu, Z. Spatio-temporal Graph Convolutional Networks: A Deep Learning Framework for Traffic Forecasting. *arXiv* **2017**, arXiv:1709.04875.

41. Zhao, L.; Song, Y.; Zhang, C.; Liu, Y.; Wang, P.; Lin, T.; Deng, M.; Li, H. T-GCN: A Temporal Graph Convolutional Network for Traffic Prediction. *IEEE Trans. Intell. Transp. Syst.* **2019**, *21*, 3848–3858. [CrossRef]

 *electronics*

Article

The Detection of Black Ice Accidents for Preventative Automated Vehicles Using Convolutional Neural Networks

Hojun Lee [1], Minhee Kang [2], Jaein Song [3] and Keeyeon Hwang [1,*]

[1]   Department of Urban Design & Planning, Hongik University, Seoul 04066, Korea; b513080@naver.com
[2]   Department of Smartcity, Hongik University Graduate School, Seoul 04066, Korea; speakingbee@hanmail.net
[3]   Research Institute of Science and Technology, Hongik University, Seoul 04066, Korea; wodlsthd@nate.com
*   Correspondence: keith@hongik.ac.kr; Tel.: +82-010-8654-7415

Received: 13 November 2020; Accepted: 16 December 2020; Published: 18 December 2020

Abstract: Automated Vehicles (AVs) are expected to dramatically reduce traffic accidents that have occurred when using human driving vehicles (HVs). However, despite the rapid development of AVs, accidents involving AVs can occur even in ideal situations. Therefore, in order to enhance their safety, "preventive design" for accidents is continuously required. Accordingly, the "preventive design" that prevents accidents in advance is continuously required to enhance the safety of AVs. Specially, black ice with characteristics that are difficult to identify with the naked eye—the main cause of major accidents in winter vehicles—is expected to cause serious injuries in the era of AVs, and measures are needed to prevent them. Therefore, this study presents a Convolutional Neural Network (CNN)-based black ice detection plan to prevent traffic accidents of AVs caused by black ice. Due to the characteristic of black ice that is formed only in a certain environment, we augmented image data and learned road environment images. Tests showed that the proposed CNN model detected black ice with 96% accuracy and reproducibility. It is expected that the CNN model for black ice detection proposed in this study will contribute to improving the safety of AVs and prevent black ice accidents in advance.

Keywords: automated vehicle; black ice; CNN; traffic accidents; prevention

---

1. Introduction

As discussions on the fourth industrial revolution become more active, there is a movement to utilize big data, artificial intelligence, and 5G. Among them, Automated Vehicles (AVs), a collection of various technologies, are attracting attention in the transportation field. AVs are expected to bring effects such as improving mobility for the vulnerable and reducing traffic congestion costs and are expected to minimize human and material losses in terms of preventing traffic accidents caused by driver negligence [1,2]. Currently, various companies such as Google, NVIDIA, and Tesla are developing and experimenting with AV systems, and each country is reorganizing its institutional foundation to prepare for the commercialization of AVs. Despite these efforts, however, traffic accidents continue to occur in autonomous driving situations, and the social acceptability of AVs has emerged due to Uber's pedestrian deaths in 2018 [3–5]. In order to solve these problems fundamentally, Germany and the United States have issued an ethics guideline for AVs [6,7]. The guidelines specify the need to develop principles to cope with dilemma situations, along with information on the preventive design of the AVs to avoid accidents. Preventive design of AVs is an issue about risk management that can occur in a realistic driving environment, changing from passive safety systems to active safety systems research [8]. In addition, recently, there has been a change in the tendency toward preventing accidents themselves by learning all the accident situations related to AVs [9]. While various preventive design studies are being carried out, there is a lack of research on preventing black ice accidents, which are the main cause of large-scale traffic accidents in winter. Black ice is a thin ice film formed on the road by combining rain and snow with pollutants such as dust, which is likely to lead to fatal accidents because it is difficult to identify with the naked eye. As black ice is considered to be a potential accident factor even in the era of commercialization of AVs, it is expected that technologies that can detect it in advance, and thus prevent accidents, will be required. Therefore, we will adopt the Convolutional Neural Network method, which is known to detect object's images most effectively, to present measures for preventing AV-related black ice accidents in the study.

This study is conducted in the following order: Section 2 discusses the research on the use of Convolutional Neural Networks (CNN) in the field of transportation and derives the differentiation of this research, while Seciton 3 sets up the CNN model learning environment for the detection of black ice. Seciton 4 identifies and analyzes learning results through models, and Seciton 5 presents implications and future studies with a brief summary.

2. Literature Review

In this chapter, we consider existing detection methods for black ice and studies using CNN in the transportation field to derive the differentiation of this study.

2.1. Black Ice Detection Methods

Methods for detecting black ice include sensors [10–12], sound waves [13,14], and light sources [15]. Habib Tabatabai et al. (2017) [10] conducted a study to detect black ice, ice and water in roads and bridges using sensors embedded in concrete. In this study, a sensor that detects the road surface condition was proposed through the change of electrical resistance between stainless steel columns inside concrete. As a result of conducting experiments under various surface conditions, it was suggested that the proposed sensor can effectively detect the road condition, thereby preventing various accidents. Nuerasimuguli ALIMASI et al. (2012) [11] conducted a study to develop a black ice detector consisting of an optical sensor and an infrared thermometer. The study was conducted in Route 39 around Sekihoku Pass, Hokkaido, and was conducted on a total of six road conditions (dry, wet, "sherbet", compact snow, glossy compacted snow, black ice) and diffuse reflection and reflection. As a result of the experiment, black ice had a large specular reflection (R_s) and a small diffuse reflection (R_D) was measured, resulting in a low (R_s/R_D) value. Youngis E. Abdalla et al. (2017) [12] proposed a system for detecting black ice using Kinect. The types of ice (Soft Ice, Wet Snow, Hard Ice, Black Ice) were classified and the thickness and volume of the ice were measured using Kinect. Experiments have shown that the types of ice formed in the range of 0.82 m to 1.52 m from the camera can be distinguished, and the error rate of measured thickness and volume is very low, suggesting that black ice can be detected by utilizing Kinect. Xinxu Ma et al. (2020) [15] studied a black ice detection method using a three-wavelength non-contact optical technology. The study conducted an experiment to distinguish dry, wet, black ice, ice and snowy conditions using three wavelengths (1310 nm, 1430 nm, 1550 nm). As a result of the experiment, it was confirmed that black ice was detected through the reflectance of each wavelength, and it was suggested that it can be used as basic data for the development of equipment to detect road conditions.

2.2. Deep Learning Applications to Intelligent Transportation

Artificial Intelligence (AI) methodologies are currently being used in various fields, and CNN studies using image data for the detection of vehicles and pedestrians, detection of traffic signs, and detection of road surface are actively carried out in the transportation field.

First of all, for vehicle and pedestrian detection, studies using AlexNet [16,17], VGG (Visual Geometry Group) 16 [18], Mask R-CNN [19], and Faster R-CNN [20,21] existed, and a comparative analysis of the performance of Faster R-CNN and YOLO (You Only Look Once) [22,23]. Lele Xie et al. (2018) [24] conducted a vehicle license plate detection study at various angles using CNN-based

Multi-Directive YOLO (MD-YOLO). The study proposed an ALMD-YOLO structure combining CNN and MD-YOLO, and compared the performance of various models (ALMD-YOLO, Faster R-CNN, SSD (Single Shot multibox Detector), MD-YOLO, etc.) and found that the newly proposed ALMD-YOLO had the best performance. It also suggested that the simple structure of the model reduced the computational time and that a high-performance multi-way license plate detection model could be established. Ye Yu et al. (2018) [25] proposed a CNN-based Feature Fusion based Car Model Classification Net (FF-CMNET) for the precise classification of vehicle models. The above study utilized FF-CMNET, which combines UpNet to extract the upper features of the car's frontal image and DownNet to extract the lower features. Experiments have shown performance better than traditional CNN methodologies (AlexNet, GoogLeNet, and Network in Network (NIN)) in terms of extracting the car's fine features. M. H. Putra et al. (2018) [26] conducted a study using YOLO to detect people and cars. Unlike the traditional YOLO structure, the above study proposed a modified YOLO structure using seven convolutional layers and compared their performance. As a result of the study, the modified YOLO's 11×11 grid cells model had a lower mAP compared to the traditional YOLO model, but had better processing speed. In addition, tests with actual images showed that small-sized people and cars could be extracted.

Second, in the case of traffic sign detection, there have been many studies using the basic CNN structure [27–30], Mask R-CNN [31,32], and Faster R-CNN [33,34]. Rongqiang Qian et al. (2016) [35] conducted a study using Fast R-CNN to recognize traffic signs on road surfaces. To enhance the performance of the model, the experiment was conducted by utilizing MDERs (Maximally Stable Extremal Regions) and EdgeBoxes algorithms in the object recognition process. The results of the experiment showed that the Recall rate was improved, with an average precision of 85.58%. Alexander Shustanov and Yakimov, P. (2017) [36] conducted a CNN model design study for real-time traffic sign recognition. The study used modified Generalized Hough Transform (GHT) and CNN, with 99.94% accuracy. It was also confirmed that the proposed algorithm could process high-definition images in real time and accurately recognize traffic signs farther away than similar traffic sign recognition systems. Lee, H.S. and Kim, K. (2018) [37] conducted a study using CNN to recognize the boundaries of traffic signs. They designed CNN based on SSD architecture, and unlike previous studies, they proposed a method of estimating the positions of signs and converting them into boundary estimates. Experiments have confirmed that various types of traffic sign boundaries can be detected quickly.

Finally, studies of road surface detection were reviewed to identify road surface conditions and to detect road cracks. Juan Carrillo et al. (2020) [38] and Guangyuan Pan et al. (2020) [39] are both studies that identify road surface conditions. The above studies divided the data into three and four classes and compared the performance of the CNN model. Studies found that up to 91% accuracy was derived, and CNN showed excellent performance in road surface identification. In the road crack detection study, Janpreet Singh et al. (2018) [40] conducted a study using Mask R-CNN to detect road damage in images taken with smartphones. The data utilized 9053 road damage images taken with smartphones, and the CNN's structure utilized Mask R-CNN. Experiments have confirmed that road damage is detected effectively, showing high accuracy and 0.1 s processing speed. Zheng Tong et al. (2018) [41] conducted a study to classify the length of asphalt cracks using DCNN (Deep Convolutional neural networks). Data collection was conducted in various places and weather conditions, and the data were divided into eight classes from 0 cm to 8 cm in 1 cm increments. As a result of the experiment, the accuracy was 94.36% and the maximum length error was 1 cm, and it was suggested that the length of the crack can be classified as well as the existence of a simple crack. Baoxian Li et al. (2020) [42] conducted a study using CNN to classify road crack types. Road cracks were classified into a total of five types (non-crack, transverse crack, longitude crack, block crack, alligator crack) and four models were designed using the basic CNN structure. As a result of the experiment, the accuracy of the four models designed was 94% or more. It was also confirmed that CNN, which has a 7×7 size of the reactive field, was the best choice for crack detection.

2.3. Summary

In summary, it was confirmed that various studies using black ice detection research and CNN in the transportation field are in progress. In addition, a study on CNN in the transportation sector has been conducted to detect the most important objects that make up road environment, such as pedestrians, vehicles, traffic signs, and road surfaces, and object detection using CNN shows a fast processing speed and high accuracy. In spite of such studies, it is expected that there is a limit to preventing black ice accidents in advance due to problems such as the installation of traditional black ice detection systems. Accordingly, this study proposes a method to detect black ice by identifying road conditions based on the CNN technique to prevent black ice accidents in AVs

3. Learning Environment Setting

CNN is a type of AI that uses convolutional computation, which emerged in 1998 when Yann LeCun proposed the LeNet-5 [43]. CNN is one of the most popular methodologies in image analysis, with features that maintain and deliver spatial information on images by adding synthetic and pooling layers to existing Artificial Neural Networks (ANN) to understand dimensional characteristics. As we considered earlier, there are various studies using CNN in the transportation sector, but the study of black ice detection on the road has only thus far been conducted using other methodologies (sensors and optics) [10–15] other than research using AI. Black ice is reckoned to be a potential accident factor in the future era of AVs as it leads to large-scale collisions in winter due to features that are hard to distinguish with the naked eye. Accordingly, we will perform the detection of black ice by utilizing the CNN technique, which is considered to have excellent performance in object detection using images, rather than the traditional black ice detection methods.

The proposed learning environment of the CNN model for black ice detection consists largely of data collection and preprocessing, model design and the learning process. In this chapter, we set up the data collection, 1st preprocessing, and 2nd preprocessing, and the model was designed and learning undertaken (see Figure 1).

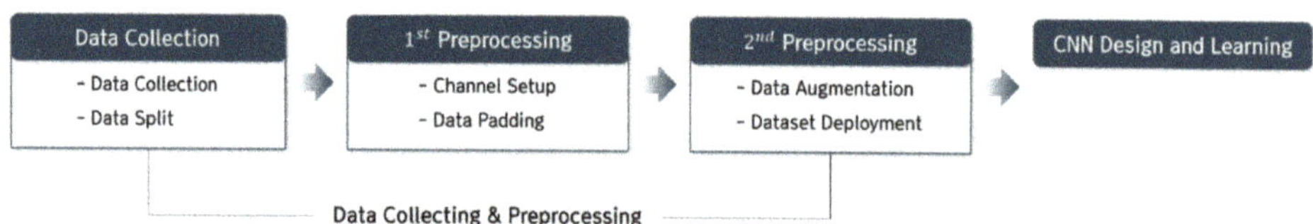

Figure 1. Learning Environment Setting Process.

3.1. Data Collecting and Preprocessing

This chapter consists of data collection for learning black ice detection, 1st preprocessing, and 2nd preprocessing.

3.1.1. Data Collection

In data collection, the method of data collection and the splits on the collected data were performed.

1. Data Collection

For learning, image data was collected using Google Image Search, and data was collected in four categories: road, wet road, snow road and black ice. During the collection process, image data taken in various regions and road environments were obtained, and a total of 2230 image data were collected as shown in Table 1.

Table 1. The Number of Image Data.

	Road	Wet Road	Snow Road	Black Ice	Total
Number	730	610	570	320	2230

2. Data Split

We split various regional and road environment data collected through Google Image Search. This is the process of removing objects that interfere with the extraction of features, such as road structures, lanes, and shoulder, within the image so that the characteristics of each category can be clearly identified. In this process, there were pros and cons depending on the size of the data (see Table 2). Smaller data sizes have disadvantages in identifying image characteristics compared to larger cases, but they have the advantages of having a large number of images and having deep neural network implementations. On the other hand, when the data are large, feature extraction can be more accurate because it can clearly identify the characteristics of the image compared to the smaller image, but the disadvantage is that the number of images is reduced and the deep neural network is difficult to implement. Accordingly, in this study, data crop was carried out in 128 × 128 px size to proceed with learning through deep neural network and large number of images. The results of the data split are shown in Figure 2.

Table 2. (Dis) advantage by data size.

	256 × 256 px	128 × 128 px
Advantage	Easy to identify image characteristics	Large number of images Deep neural network can be implemented
Disadvantage	Small number of images Unable to implement deep neural network	Hard to identify image characteristics

Figure 2. The results of the data split.

3.1.2. 1st Preprocessing

In the 1st Preprocess, the channel setup and data padding for learning were performed.

1. Channel Setup

The color image of 128 × 128 px obtained earlier through data split has the advantage of being easy to identify the characteristics of the data in the form of three channels. However, since there are three channels of data, the size of the data is large, which limits the number of learning data and the implementation of deep neural networks, this study has transformed the data into black and white image data to conduct learning (see Table 3).

Table 3. Features of RGB, GRAYSCALE.

	RGB	**GRAYSCALE (Black and White)**
Number of Channels	3 Channels	1 Channel
Feature	Large data size	Small data size
Advantage	Easy to identify image characteristics	No limit on the number of learning data Deep neural networks can be implemented
Disadvantage	Limited number of learning data Deep neural network impossible to implement	Hard to identify image characteristics

2.   Data Padding

Data padding is one of the ways to resize learning images by adding spaces and meaningless symbols to the end of existing data. As a result of learning without data padding in the augmentation conducted during the 2nd preprocess of this study, very low accuracy (25%) and high loss values were identified (Table 4). This is because the edges of the image data are distorted by the data enhancement. Accordingly, in this study, the image data were padded to prevent distortion of the edges of the data.

Table 4. Data Augmentation and Learning Results.

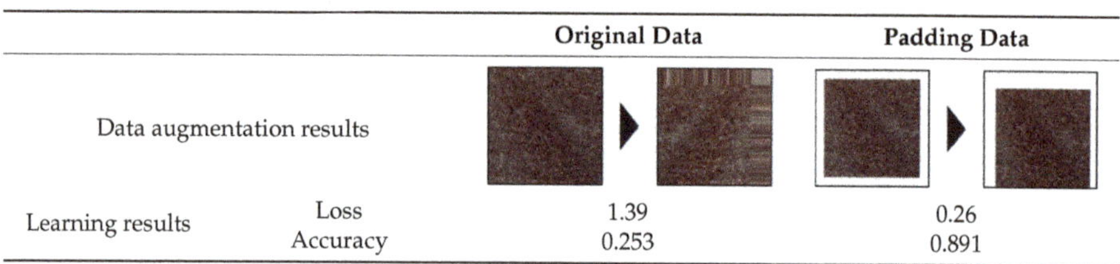

		Original Data	**Padding Data**
Data augmentation results			
Learning results	Loss	1.39	0.26
	Accuracy	0.253	0.891

During the 1st preprocessing, during which channel setup and data padding were performed, image data of 150 × 150 px in GRAYSCALE format were obtained as follows: There are 4900 road and wet road image data and 4900 snow road and black ice image data (Table 5).

Table 5. Number of data through 1st preprocessing.

Class	Size	Number
Road		4900
Wet road	150 × 150 px	4900
Snow road		3900
Black ice		3900
Total		17,600

3.1.3. 2nd Preprocessing

In the 2nd preprocessing, there was a limit to collecting more diverse image data through Google Image Search, so data sets were built through data augmentation to increase their accuracy.

For AI learning models, large amounts of data are essential for high accuracy and prevention of overfitting [44]. In particular, black ice, which is intended to be detected in this study, is characterized by seasonal characteristics and unusual forming conditions that do not occur in many places. As a result, the data collection process did not collect a large amount of data compared to the other data. Accordingly, to improve the accuracy of CNN proposed in this study, the ImageDataGenerator function provided by the Keras library [45] was used to augment the data under the conditions in (Table 6).

Table 6. Setting Data Augmentation Value.

Transformation Type	Value
Rotation	20
Width shift	0.15
Height shift	0.15
Zoom	0.1

The process of building a data set through data augmentation is as follows. From the previously obtained 17,600 sheets of data, 1000 were randomly extracted per class and set as test data. Since then, the data have been augmented using the ImageDataGenerator function for the rest data, which has built 10,000 data per class. The train data and validation data were then set at 8:2. Accordingly, the final data set was set to 8:2:1 in all classes to proceed with the learning (Figure 3 and Table 7).

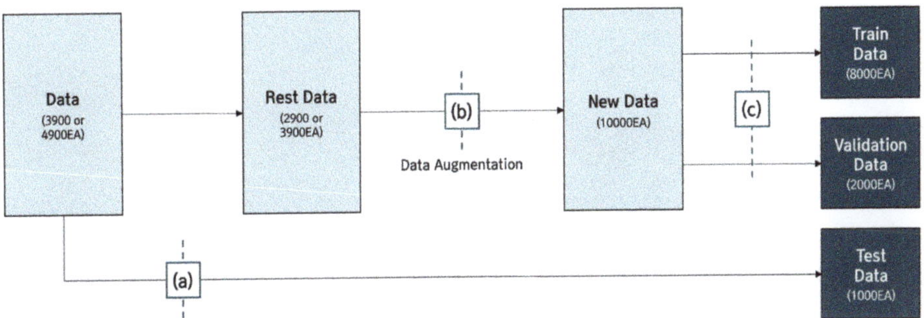

Figure 3. (**a**) Numbers of 1000 data are randomly selected and set as test data; (**b**) data augmentation with ImageDataGenerator functions; (**c**) train data and validation data set at 8:2 ratio.

Table 7. Dataset for class.

Class	Train Data	Validation Data	Test Data	Total
Road				
Wet road	8000	2000	1000	11,000
Snow road				
Black ice				

3.2. CNN Design and Learning

The structure of the CNN model used in this study consists of the Feature Extraction and Classification as shown in Figure 4. In the feature extraction, two convolutional layers, two max-pooling layers, and one dropout layer were arranged to conduct two iterations (Figure 4a). Each layer was then arranged once and repeated twice (Figure 4b). In addition, we used ReLU (Rectified Linear Unit), which has a fast learning speed and prevents gradient vanishing [46], as the activation function. In the case of kernel size of the convolutional layer, (3,3) was applied because the repetition of (3,3) has a fast learning speed and extracts features well [47]. The Stride of the max pooling layer was (2,2), and the Dropout rate of the dropout layer was experimentally applied 0.2. In the classification, the Fully-Connected layer and the Dropout layer were alternately placed, Softmax was applied to the output layer (Figure 4c), and the SGD (Stochastic Gradient Descent) Optimizer was used for high experimental accuracy. In addition, we applied 200 epochs, 32 batch size and the earlystopping function for optimizing the model and preventing overfitting. The earlystopping function terminates learning when there is no room for improvement in the model. Therefore, we designed this study to stop learning if the validation loss does not update the minimum value within 20 times (see Table 8)

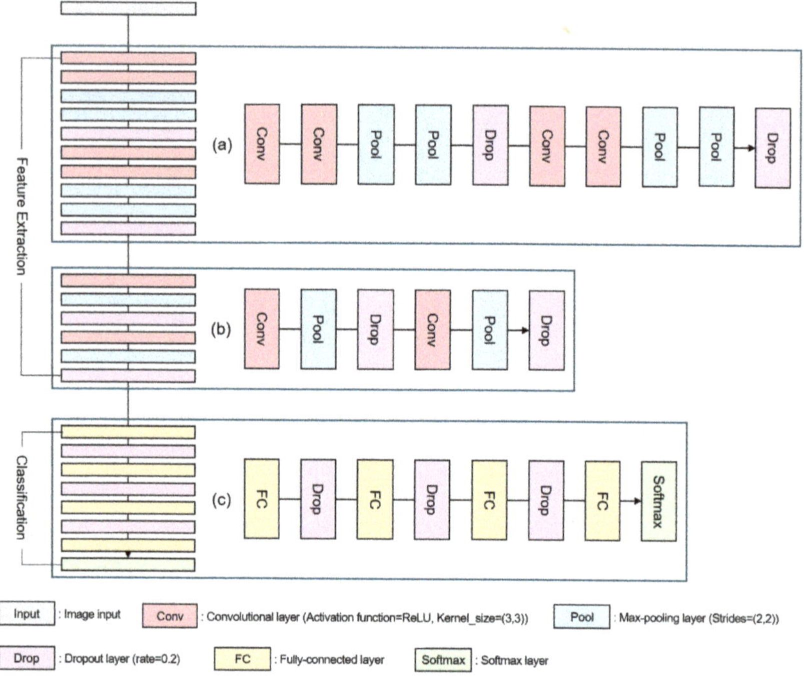

Figure 4. CNN model structure; (**a**) after placing the convolutional layer and the max-pooling layer twice, place the dropout layer once and repeat training twice; (**b**) place the convolutional layer, the max-pooling layer and dropout layer once and repeat training twice; (c) fully-connected and dropout layers are alternately placed and softmax is applied to the output layer.

Table 8. CNN Model Setup Value.

Class	Value
Activation Function	ReLU
Kernel size	(3,3)
Strides	(2,2)
Dropout rate	0.2
Optimizer	SGD
Epoch	200
Batch size	32
Earlystopping	20

4. Result

4.1. Result

Based on the CNN model, the loss of train data and test data was found to be 0.008 and 0.097, respectively, and the accuracy was 0.998 and 0.982, respectively (see Figure 5 and Table 9).

For more detailed learning results analysis, performance indicators by class were identified with train data and test data. First, we analyzed a confusion matrix for train data and checked the classification results by class. Confusion matrix is a matrix for comparing the predicted class with the actual class to measure the prediction performance through training, with the x-axis representing the predicted class and the y-axis representing the actual class. The results of the Confusion matrix

(Figure 6) showed that some mutual confusion occurred between black ice and snow road, and when the actual class was wet road, it was predicted as road.

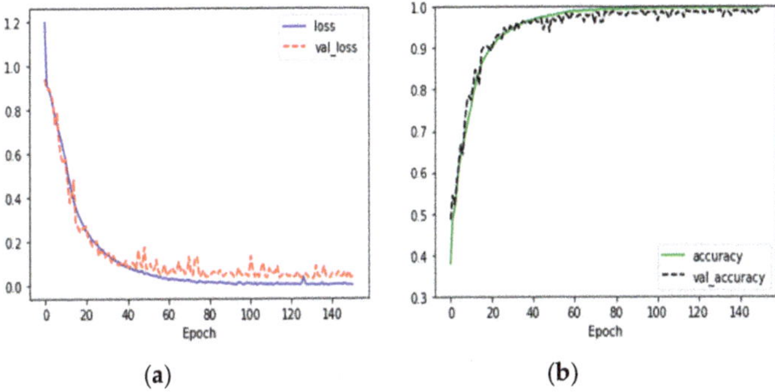

(a)   **(b)**

Figure 5. Training result: (**a**) The *x*-axis of the left graph represents the value of Epoch, the *y*-axis represents the value of loss; (**b**) the *x*-axis of the right graph represents Epoch, and the *y*-axis represents accuracy.

Table 9. Training Result.

Class	Loss	Accuracy
Train	0.008	0.998
Test	0.097	0.982

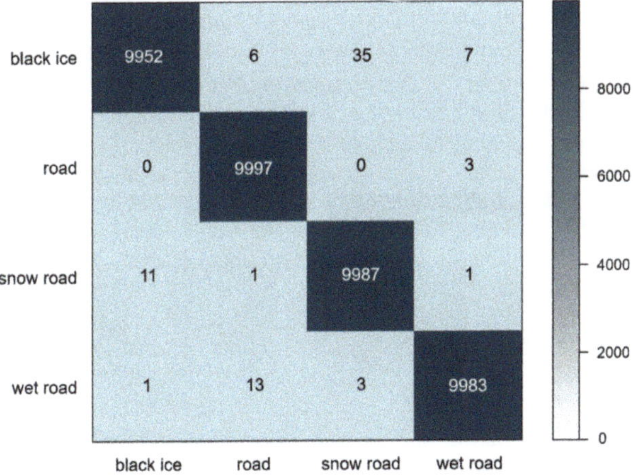

Figure 6. Confusion matrix; A matrix written to measure the prediction performance through training, the *x*-axis represents the predicted class and the *y*-axis represents the actual class. The results showed that (x,y) = (snow road, black ice) = 35, (black ice, snow road) = 11, (road, wet road) = 13.

Secondly, the calculation and analysis of accuracy, precision, and recall of each class was conducted on test data. The calculation results of each performance indicator are shown in Table 10. This shows that the accuracy of black ice, wet road and snow road is measured as relatively low, which is estimated to be the result of loss of light characteristics in the same way as the Confusion matrix analyzed earlier. However, the average values of accuracy, precision and call were 0.982, 0.983, 0.983, and 0.983, which are considered to have produced significant learning outcomes, even though the data are not relevant to learning.

Table 10. Accuracy, precision, recall results by class.

Class	Accuracy	Precision	Recall
Road	0.996	0.99	1.00
Wet road	0.989	0.99	0.99
Snow road	0.981	0.97	0.98
Black ice	0.961	0.98	0.96
Average	0.982	0.983	0.983

4.2. Discussion

The 69-car collision on the Virginia Expressway in December 2019 and the traffic collision on the Yeongcheon Expressway in Korea in December of the same year were all caused by black ice. As such, casualties from black ice accidents continue to occur worldwide. To prevent this, measures such as installing grooving (a construction method that makes small grooves on the surface of the road to reduce braking distance), installing LED signs, and installing heat wires on the road are currently being proposed. However, since the proposed measures are not a preliminary measure to prevent accidents, a proactive strategy to prevent black ice accidents is being devised. Accordingly, the study was conducted to detect black ice in advance using AI methodologies to extract high accuracy. As further confirmed earlier, the existing black ice detection method is expected to present a prevention measure using light [10,15], as a methodology using light sources is used. This strategy is due to the light reflection feature of black ice, and the image data used in this study also needs to be considered. As confirmed by the analysis results, we have checked that the black ice and snow road appear in the Confusion matrix, and we will explain the cause through the train data of two categories. Figure 7 shows the RGB version of train data, with black ice data showing reflections of light, and snow road data showing snow crystals and slush. However, in this study, because the neural network was designed low due to the limitations of computing and the transformation to GRAYSCALE was carried out (Figure 8), it was difficult to clearly identify these characteristics, resulting in confusion between the two classes and relatively low black ice accuracy of the test data.

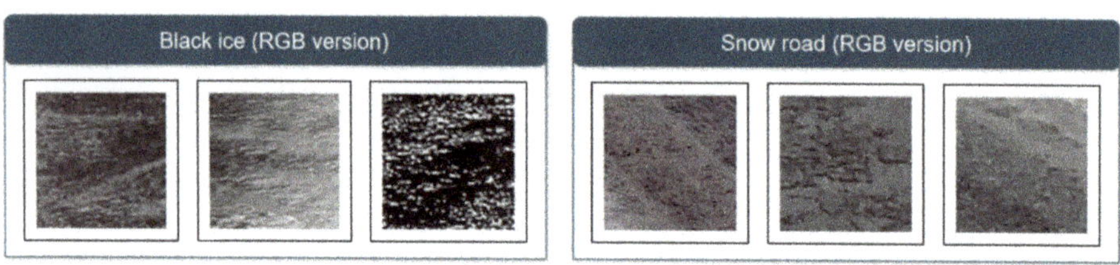

Figure 7. RGB version of train data (black ice and snow road); in the case of black ice, it can be seen that it shimmers due to light reflection, and in the case of snow data, snow crystals and slush appear.

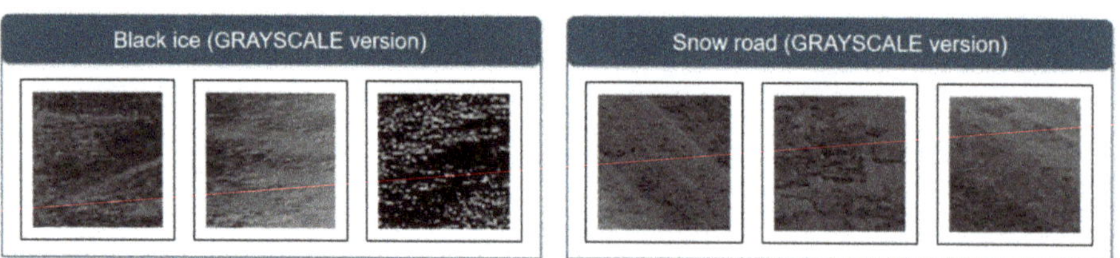

Figure 8. GRAYSCALE version of train data (black ice and snow road); relative to the data in the previous figure, the reflection of black ice and snow crystals and slush are not clearly identified.

4.3. Application Method

We propose an application that can take advantage of the CNN-based black ice detection method we proposed. It will offer a means of preventing accidents by being mounted on AVs and CCTVs. First of all, it is expected that the camera attached to the vehicle will be able to detect black ice in advance by pre-learning the detection of black ice in accordance with the engineering characteristics of AVs. In addition, CCTV is expected to become an important medium for C-ITS (Cooperative Intelligent Transport System) in the future, so it is expected that it can be installed in a proper black ice accident prediction area to prevent accidents. CCTV and cameras installed in AVs are used to forward information on the presence or absence of black ice to nearby vehicles and take measures to undertake a detour of the area or reduce vehicle speeds to prevent major accidents caused by black ice.

5. Conclusions

This study conducted a study using a CNN to detect black ice that is difficult to judge visually in order to prevent black ice accidents in AVs. Data were collected via classification into four classes, and each class's train, validation, and test data were set through pre-processing of split, padding, and augmentation. Unlike the DCNN model, the CNN model proposed in this study was designed to be relatively simple but showed an excellent performance with an accuracy of about 96%. This suggests that it is more effective to optimize the neural network depth according to the object to be detected rather than to detect black ice by increasing the amount of computation through a complex neural network model. In addition, in this study, a neural network was designed and learned through GRAYSCALE as a feature of black ice mainly formed at dawn, but it was found that some specific classes were confused due to the loss of light characteristics. Accordingly, we plan to conduct research on neural network design that is more optimized for black ice detection by utilizing RGB images in the future. Additionally, since the data were collected through Google Image Search, only images detected close to the object are classified. Accordingly, we plan to construct a CNN model applicable to various situations by setting the distance and angle to the object to be detected in various ways [48–50] in the future.

This study is significant in that black ice, which is deemed a potential risk factor even in the era of AVs, was detected using AI, not sensors and wavelengths. It is expected that this will prevent black ice accidents of AVs and will be used as basic data for future convergence research.

Author Contributions: Conceptualization, M.K. and K.H.; Data curation, H.L.; Formal analysis, H.L.; Methodology, M.K.; Project administration, M.K. and J.S.; Software, H.L.; Supervision, K.H.; Visualization, H.L.; Writing—original draft, H.L., M.K. and K.H.; Writing—review and editing, M.K. and K.H. All authors have read and agreed to the published version of the manuscript.

Funding: This research was supported by Basic Science Research Program through the National Research Foundation of Korea (NRF) funded by the Ministry of Education (No. 2020R1F1A106988411).

References

1. Lee, K.; Jeon, S.; Kim, H.; Kum, D. Optimal path tracking control of autonomous vehicle: Adaptive full-state linear quadratic gaussian (lqg) control. *IEEE Access* **2019**, *7*, 109120–109133. [CrossRef]
2. Singh, S. *Critical Reasons for Crashes Investigated in the National Motor Vehicle Crash Causation Survey (No. DOT HS 812 115)*; NHTSA's National Center for Statistics and Analysis: Washington, DC, USA, 2015.
3. Federal Ministry of Transport and Digital Infrastructure. *Ethics Commission: Automated and Connected Driving*; Federal Ministry of Transport and Digital Infrastructure: Berlin, Germany, 2017.
4. National Highway Traffic Safety Administration. *Federal Automated Vehicles Policy: Accelerating the Next Revolution in Roadway Safety*; National Highway Traffic Safety Administration: Washington, DC, USA, 2016.
5. Abraham, H.; Lee, C.; Brady, S.; Fitzgerald, C.; Mehler, B.; Reimer, B.; Coughlin, J.F. Autonomous vehicles and alternatives to driving: Trust, preferences, and effects of age. In Proceedings of the Transportation Research Board 96th Annual Meeting, Washington, DC, USA, 8–12 January 2017.

6. Zhang, T.; Tao, D.; Qu, X.; Zhang, X.; Lin, R.; Zhang, W. The roles of initial trust and perceived risk in public's acceptance of automated vehicles. *Transp. Res. Part C Emerg. Technol.* **2019**, *98*, 207–220. [CrossRef]

7. Hartwich, F.; Witzlack, C.; Beggiato, M.; Krems, J.F. The first impression counts–A combined driving simulator and test track study on the development of trust and acceptance of highly automated driving. *Transp. Res. Part F Traffic Psychol. Behav.* **2019**, *65*, 522–535. [CrossRef]

8. Kim, K.; Kim, B.; Lee, K.; Ko, B.; Yi, K. Design of integrated risk management-based dynamic driving control of automated vehicles. *IEEE Intell. Transp. Syst. Mag.* **2017**, *9*, 57–73. [CrossRef]

9. The New York Times Online. Available online: https://www.nytimes.com/2020/10/26/technology/driverless-cars.html (accessed on 3 December 2020).

10. Tabatabai, H.; Aljuboori, M. A novel concrete-based sensor for detection of ice and water on roads and bridges. *Sensors* **2017**, *17*, 2912. [CrossRef]

11. Alimasi, N.; Takahashi, S.; Enomoto, H. Development of a mobile optical system to detect road-freezing conditions. *Bull. Glaciol. Res.* **2012**, *30*, 41–51. [CrossRef]

12. Abdalla, Y.E.; Iqbal, M.T.; Shehata, M. Black Ice detection system using Kinect. In Proceedings of the IEEE 30th Canadian Conference on Electrical and Computer Engineering (CCECE), Windsor, ON, Canada, 30 April–3 May 2017; pp. 1–4.

13. Minullin, R.G.; Mustafin, R.G.; Piskovatskii, Y.V.; Vedernikov, S.G.; Lavrent'ev, I.S. A detection technique for black ice and frost depositions on wires of a power transmission line by location sounding. *Russ. Electr. Eng.* **2011**, *82*, 541–543. [CrossRef]

14. Gailius, D.; Jačėnas, S. Ice detection on a road by analyzing tire to road friction ultrasonic noise. *Ultragarsas Ultrasound* **2007**, *62*, 17–20.

15. Ma, X.; Ruan, C. Method for black ice detection on roads using tri-wavelength backscattering measurements. *Appl. Opt.* **2020**, *59*, 7242–7246. [CrossRef]

16. Han, J.; Koo, B.; Choi, K. Obstacle detection and recognition system for self-driving cars. *Converg. Inf. Pap.* **2017**, *7*, 229–235. (In Korean)

17. Gao, H.; Cheng, B.; Wang, J.; Li, K.; Zhao, J.; Li, D. Object classification using CNN-based fusion of vision and LIDAR in autonomous vehicle environment. *IEEE Trans. Ind. Inform.* **2018**, *14*, 4224–4231. [CrossRef]

18. Ammour, N.; Alhichri, H.; Bazi, Y.; Benjdira, B.; Alajlan, N.; Zuair, M. Deep learning approach for car detection in UAV imagery. *Remote Sens.* **2017**, *9*, 312. [CrossRef]

19. Nafi'i, M.W.; Yuniarno, E.M.; Affandi, A. Vehicle Brands and Types Detection Using Mask R-CNN. In Proceedings of the International Seminar on Intelligent Technology and Its Applications (ISITIA), Surabaya, Indonesia, 28–29 August 2019; pp. 422–427.

20. Zhang, L.; Lin, L.; Liang, X.; He, K. Is faster R-CNN doing well for pedestrian detection? In Proceedings of the European Conference on Computer Vision, Amsterdam, The Netherlands, 11–14 October 2016; pp. 443–457.

21. Chen, Y.Y.; Jhong, S.Y.; Li, G.Y.; Chen, P.H. Thermal-based pedestrian detection using faster r-cnn and region decomposition branch. In Proceedings of the International Symposium on Intelligent Signal Processing and Communication Systems (ISPACS), Taipei, Taiwan, 3–6 December 2019; pp. 1–2.

22. Ammar, A.; Koubaa, A.; Ahmed, M.; Saad, A. Aerial images processing for car detection using convolutional neural networks: Comparison between faster r-cnn and yolov3. *arXiv* **2019**, arXiv:1910.07234.

23. Benjdira, B.; Khursheed, T.; Koubaa, A.; Ammar, A.; Ouni, K. Car detection using unmanned aerial vehicles: Comparison between faster r-cnn and yolov3. In Proceedings of the 1st International Conference on Unmanned Vehicle Systems-Oman (UVS), Muscat, Oman, 5–7 February 2019; pp. 1–6.

24. Xie, L.; Ahmad, T.; Jin, L.; Liu, Y.; Zhang, S. A new CNN-based method for multi-directional car license plate detection. *IEEE Trans. Intell. Transp. Syst.* **2018**, *19*, 507–517. [CrossRef]

25. Yu, Y.; Jin, Q.; Chen, C.W. FF-CMnet: A CNN-based model for fine-grained classification of car models based on feature fusion. In Proceedings of the IEEE International Conference on Multimedia and Expo (ICME), San Diego, CA, USA, 23–27 July 2018; pp. 1–6.

26. Putra, M.H.; Yussof, Z.M.; Lim, K.C.; Salim, S.I. Convolutional neural network for person and car detection using yolo framework. *J. Telecommun. Electron. Comput. Eng.* **2018**, *10*, 67–71.

27. Zhu, Z.; Liang, D.; Zhang, S.; Huang, X.; Li, B.; Hu, S. Traffic-sign detection and classification in the wild. In Proceedings of the IEEE Conference on Computer Vision and Pattern Recognition, Las Vegas, NV, USA, 27–30 June 2016; pp. 2110–2118.

28. Vennelakanti, A.; Shreya, S.; Rajendran, R.; Sarkar, D.; Muddegowda, D.; Hanagal, P. Traffic sign detection and recognition using a cnn ensemble. In Proceedings of the IEEE International Conference on Consumer Electronics (ICCE), Las Vegas, NV, USA, 11–13 January 2019; pp. 1–4.

29. Alghmgham, D.A.; Latif, G.; Alghazo, J.; Alzubaidi, L. Autonomous traffic sign (ATSR) detection and recognition using deep CNN. *Procedia Comput. Sci.* **2019**, *163*, 266–274. [CrossRef]

30. Peemen, M.; Mesman, B.; Corporaal, H. Speed sign detection and recognition by convolutional neural networks. In Proceedings of the 8th International Automotive Congress, Eindhoven, The Netherlands, 16–17 May 2011; pp. 162–170.

31. Malbog, M.A. MASK R-CNN for Pedestrian Crosswalk Detection and Instance Segmentation. In Proceedings of the IEEE 6th International Conference on Engineering Technologies and Applied Sciences (ICETAS), Kuala Lumpur, Malaysia, 20–21 December 2019; pp. 1–5.

32. Tabernik, D.; Skočaj, D. Deep learning for large-scale traffic-sign detection and recognition. *IEEE Trans. Intell. Transp. Syst.* **2019**, *21*, 1427–1440. [CrossRef]

33. Kukreja, R.; Rinchen, S.; Vaidya, B.; Mouftah, H.T. Evaluating Traffic Signs Detection using Faster R-CNN for Autonomous driving. In Proceedings of the IEEE 25th International Workshop on Computer Aided Modeling and Design of Communication Links and Networks (CAMAD), Pisa, Italy, 14–16 September 2020; pp. 1–6.

34. Boujemaa, K.S.; Berrada, I.; Bouhoute, A.; Boubouh, K. Traffic sign recognition using convolutional neural networks. In Proceedings of the International Conference on Wireless Networks and Mobile Communications (WINCOM), Rabat, Morocco, 1–4 November 2017; pp. 1–6.

35. Qian, R.; Liu, Q.; Yue, Y.; Coenen, F.; Zhang, B. Road surface traffic sign detection with hybrid region proposal and fast R-CNN. In Proceedings of the 12th International Conference on Natural Computation, Fuzzy Systems and Knowledge Discovery (ICNC-FSKD), Changsha, China, 13–15 August 2016; pp. 555–559.

36. Shustanov, A.; Yakimov, P. CNN design for real-time traffic sign recognition. *Procedia Eng.* **2017**, *201*, 718–725. [CrossRef]

37. Lee, H.S.; Kim, K. Simultaneous traffic sign detection and boundary estimation using convolutional neural network. *IEEE Trans. Intell. Transp. Syst.* **2018**, *19*, 1652–1663. [CrossRef]

38. Carrillo, J.; Crowley, M.; Pan, G.; Fu, L. Design of Efficient Deep Learning models for Determining Road Surface Condition from Roadside Camera Images and Weather Data. *arXiv* **2020**, arXiv:2009.10282.

39. Pan, G.; Muresan, M.; Yu, R.; Fu, L. Real-time Winter Road Surface Condition Monitoring Using an Improved Residual CNN. *Can. J. Civ. Eng.* **2020**. [CrossRef]

40. Singh, J.; Shekhar, S. Road damage detection and classification in smartphone captured images using mask r-cnn. *arXiv* **2018**, arXiv:1811.04535.

41. Tong, Z.; Gao, J.; Han, Z.; Wang, Z. Recognition of asphalt pavement crack length using deep convolutional neural networks. *Road Mater. Pavement Des.* **2018**, *19*, 1334–1349. [CrossRef]

42. Li, B.; Wang, K.C.; Zhang, A.; Yang, E.; Wang, G. Automatic classification of pavement crack using deep convolutional neural network. *Int. J. Pavement Eng.* **2020**, *21*, 457–463. [CrossRef]

43. LeCun, Y.; Bottou, L.; Bengio, Y.; Haffner, P. Gradient-based learning applied to document recognition. *Proc. IEEE* **1998**, *86*, 2278–2324. [CrossRef]

44. Choubisa, T.; Kashyap, M.; Chaitanya, K.K. Human Crawl vs Animal Movement and Person with Object Classifications Using CNN for Side-view Images from Camera. In Proceedings of the International Conference on Advances in Computing, Communications and Informatics (ICACCI), Bangalore, India, 19–22 September 2018; pp. 48–54.

45. Keras. Available online: https://keras.io/ (accessed on 11 November 2020).

46. Nair, V.; Hinton, G.E. Rectified linear units improve restricted boltzmann machines. In Proceedings of the 27th International Conference on International Conference on Machine Learning, Haifa, Israel, 21–24 June 2010.

47. Simonyan, K.; Zisserman, A. Very deep convolutional networks for large-scale image recognition. *arXiv* **2014**, arXiv:1409.1556.

48. Gadd, M.; Newman, P. A framework for infrastructure-free warehouse navigation. In Proceedings of the IEEE International Conference on Robotics and Automation (ICRA), Seattle, WA, USA, 26–30 May 2015; pp. 3271–3278.

49. Tourani, S.; Desai, D.; Parihar, U.S.; Garg, S.; Sarvadevabhatla, R.K.; Krishna, K.M. Early Bird: Loop Closures from Opposing Viewpoints for Perceptually-Aliased Indoor Environments. *arXiv* **2020**, arXiv:2010.01421.

50. Chen, X.; Vempati, A.S.; Beardsley, P. Streetmap-mapping and localization on ground planes using a downward facing camera. In Proceedings of the IEEE/RSJ International Conference on Intelligent Robots and Systems (IROS), Madrid, Spain, 1–5 October 2018; pp. 1672–1679.

Publisher's Note: MDPI stays neutral with regard to jurisdictional claims in published maps and institutional affiliations.

 *electronics*

Article

Estimating Micro-Level On-Road Vehicle Emissions Using the K-Means Clustering Method with GPS Big Data

Hyejung Hu [1], Gunwoo Lee [1,*], Jae Hun Kim [1] and Hyunju Shin [2]

[1]   Department of Transportation and Logistics Engineering, Hanyang University, Ansan 15588, Korea; hu20000@gmail.com (H.H.); jhkim8182@hanyang.ac.kr (J.H.K.)

[2]   Department of Smart City Engineering, Hanyang University, Seoul 04763, Korea; sinzoo@hanyang.ac.kr

*   Correspondence: gunwoo@hanyang.ac.kr; Tel.: +82-31-400-5156

Received: 15 November 2020; Accepted: 13 December 2020; Published: 15 December 2020

Abstract: Due to the advanced spatial data collection technologies, the locations of vehicles on roads are now being collected nationwide, so there is a demand for applying a micro-level emission calculation methods to estimate regional and national emissions. However, it is difficult to apply this method due to the low data collection rate and the complicated calculation procedure. To solve these problems, this study proposes a vehicle trajectory extraction method for estimating micro-level vehicle emissions using massive GPS data. We extracted vehicle trajectories from the GPS data to estimate the emission factors for each link at a specific time period. Vehicle trajectory data was divided into several groups through a k-means clustering method, in which the ratios of each operating mode were used as variables for clustering similar vehicle trajectories. The results showed that the proposed method has an acceptable accuracy in estimating emissions. Furthermore, it was also confirmed that the estimated emission factors appropriately reflected the driving characteristics of links. If the proposed method were utilized to update the link-based micro-level emission factors using continuously accumulated trajectory data for the road network, it would be possible to efficiently calculate the regional- or national-level emissions only using traffic volume.

Keywords: vehicle GPS data; driving cycle; micro-level vehicle emission estimation; link emission factors; MOVES

1. Introduction

Emissions from on-road mobile sources depend on the driving characteristics of the vehicles. The emission calculation methods are largely divided into macro-level approaches that consider the average speed of vehicles traveling along the roads as a driving characteristic and micro-level methods that reflect the change in instantaneous speed of individual vehicles as a driving characteristic [1]. Since the data on average speed, distance traveled (or link length), and traffic volume on the road section, which are necessary input data for calculating macroscopic emissions, is basically built and managed in the traffic network data, it is relatively easy to calculate the total emissions of the traffic network. For this reason, the macro-level emission estimation method has been generally applied and utilized for calculating the emissions from on-road mobile sources in different regions and countries [2–4]. However, in the macro-level emission estimation method using the average speed as the deterministic variable, the amount of vehicle emissions may be under- or overestimated because this method cannot capture the instantaneous changes in the vehicle driving speeds, such as stop-and-go traffic situations [5,6]. For this reason, there have been many claims that the micro-level emissions method should be applied in estimating regional and national emissions.

A number of previous studies have mounted GPS equipment onto only experimental vehicles to collect the vehicle trajectory data and estimate the vehicle emissions [7–9]. However, with the advancement of electronic communication technology, vehicle trajectories are being continuously collected nationwide by using vehicle navigation devices, digital tachographs (DTGs), and mobile devices, which have become available for estimating vehicle emissions [10–12]. The collection of vehicle trajectory data has become easier, and the range of spatio-temporal data collected has been expanded. As a result, it is possible to estimate micro-level emissions from the collected vehicle trajectory data.

However, there are some practical difficulties associated with adopting the micro-level approach at the regional and national levels. First, developing micro-level emission factors requires a large amount of time and cost. For vehicles with various fuel types, fleet sizes, and model years, it is necessary to conduct multiple driving tests, measure emissions under various operating conditions, and derive emission factors suitable for the micro-level emission estimation method. A fast solution is to use micro-level emission factor databases from other countries. In fact, the U.S. Environmental Protection Agency's (U.S. EPA) Motor Vehicle Emission Simulator (MOVES) [13] can estimate micro-level vehicle emissions, and it has been applied in various studies from many countries [14–18] even though the approach should be taken with caution because the classification method for vehicle types and emission standards vary by country. The second problem is related to the acquisition of vehicle trajectory data, which is required to calculate micro-level vehicle emissions. In the case of collecting the link average speed of traffic flow, the average speed data can be easily collected from traffic information systems, such as loop detectors. However, estimating vehicle emissions with micro-level emission models is limited because it requires second-by-second vehicle trajectory data. It would be ideal to collect the trajectories of all vehicles driving along the roads to estimate the emissions of local or national on-road mobile sources on a micro-level basis, but that is highly impractical. Therefore, it would be useful to apply vehicle emissions, which are estimated at a micro level, at the regional and national levels with an easier and faster method.

This study proposes a representative vehicle trajectory extraction method for estimating micro-level vehicle emissions with a limited amount of vehicle trajectory data, such as that from DTGs or mobile devices. In the method, MOVES is used for analyzing vehicle emissions at a micro level, and vehicle trajectory data is divided into several groups through a k-means clustering method, in which the ratios of each operating mode (OpMode) in MOVES are used as cluster variables for clustering similar vehicle trajectories.

The rest of this paper is organized as follows: Section 2 describes the uniqueness of the representative vehicle trajectory extraction methodology used in this study, and Section 3 explains the proposed network-level micro-level emission estimation procedure, representative vehicle trajectory extraction method, and micro-level emission factor derived from this study. Section 4 presents the results of applying the proposed method to navigation data collected in Bucheon, Gyeonggi-do in Republic of Korea. Section 5 presents the effects of using the accumulated vehicle trajectory data on the method. In Section 6, the implications learned from the analysis results and the limitations of this study are discussed. The conclusions for this study are mentioned in Section 7.

2. Literature Review

To apply the micro-level emission estimation method to on-road mobile sources, it would be ideal to collect the trajectories of all vehicles running on the road and put the data into the micro-level emission calculation program. However, doing so is highly impractical, and even if it were not, it would be inefficient because it takes a large amount of time to calculate the micro-level emissions from all driving vehicles in the transportation network. To overcome this limitation, the U.S. EPA has developed MOVES [13], and several studies have incorporated vehicle trajectories (called driving cycles) into MOVES, which represent driving characteristics by vehicle type, road type, and level of service (LOS) [19–22]. However, it is not suitable to use the MOVES driving cycle in countries other than the United States because the road geometry, vehicle type composition ratio, and driver's driving characteristics that determine the driving characteristics of a vehicle vary among countries [23,24]. Therefore, some studies conducted in countries other than the United States have applied MOVES with adjusted base emission rates to analyze vehicle emissions in those countries [19–21].

Another solution is to collect vehicle trajectory data from on-road mobile sources in the target area to be analyzed. The individual vehicle trajectories used in most studies were collected from a GPS device installed in a driving vehicle [7–9] or extracted from a microscopic traffic simulation model [25,26]. The vehicle trajectory data is collected or extracted from several vehicle types, including cars, trucks, and buses. Therefore, distinguishing the data by vehicle type is crucial for calculating vehicle emissions. Several studies have applied cluster analysis methods [27,28] to extract representative vehicle trajectories. Moreover, variables used to distinguish similar types of vehicle trajectories have included average speed, average acceleration, average deceleration, time proportion of idling mode/cruising mode/acceleration mode/deceleration mode/creeping mode, frequency of vehicle stops, mean length of driving period, average number of acceleration–deceleration changes, and root mean squared acceleration, which are aggregates representing the corresponding characteristics [29].

Previous studies associated with cluster analysis were performed to classify the vehicle trajectories into the representative driving cycles by vehicle type and road type. This study utilizes cluster analysis to assemble the vehicle trajectories of the same vehicle type on the same road section at the same time intervals into several similar vehicle trajectory groups and then uses the representative patterns of each group for estimating the corresponding highway link-based vehicle emissions. This process is expected to reflect the various driving situations that can occur in the same context. If aggregated characteristics, such as average speed and maximum acceleration, are used as variables for clustering, these measures cannot be used for micro-level emission estimation. Thus, the MOVES OpMode distribution, which can be used immediately for estimating micro-level vehicle emissions in MOVES, is applied as a variable for cluster analysis.

3. Methods

3.1. Micro-Level Emission Estimation Method Using Individual Vehicle Trajectory Data

3.1.1. Micro-Level Emission Factor

The MOVES's micro-level emission estimation methodology is applied to calculating emissions with trajectories of individual vehicles collected from vehicle navigation, odometer (i.e., DTG), and mobile devices. The MOVES database stores the basic emission factors for each pollutant at OpMode by each vehicle type, which is subdivided by fuel use, size, year of manufacture, and vehicle age. MOVES provides the result of calculating the emissions according to various conditions, such as the vehicle type composition ratio, fuel compound characteristics, and temperature and humidity on the road section.

In this study, the base emission rates for each vehicle type from the MOVES database are applied to calculate the basic emission factors at the OpModes of passenger cars, trucks, and buses, as shown in Appendixes A–C. In order to make these tables, the ratios of each subdivided vehicle type (size, fuel, vehicle age) of the target year 2017 in Bucheon, Gyeonggi-do, which is an analysis target area collected from the Korea vehicle registration information, are used as weighted values. CO, NO_x, PM_{10}, $PM_{2.5}$, and CO_2 are selected as the pollutants to be estimated, and the unit of the emission factors is g/sec. The emissions are calculated in this study without considering the adjustment for climatic conditions.

3.1.2. MOVES OpMode

Location information in the vehicle trajectory of an individual vehicle was collected from vehicle navigation and DTG data. After the speed per second is calculated with the distance from the change in vehicle location per second, the acceleration per second can be calculated. Then, the vehicle-specific power (VSP) per second is calculated with Equation (1), which is an equation for MOVES VSP [30].

The terms A, B, and C have different values for each vehicle type, and a suitable value for the corresponding vehicle type among the values presented in MOVES should be found and applied. Next, OpMode per second is found among 23 OpModes according to the VSP range and speed range (see Appendixes A–C):

$$P_{V,t} = \frac{Av_t + Bv_t^2 + Cv_t^3 + mv_t(a_t + gsin\theta_t)}{m} \tag{1}$$

where $P_{V,t}$: VSP (kW/ton) for vehicle V at time t, t: time (s), v_t: speed (m/s^2), a_t: acceleration (m/s^2), m: weight (ton), A: rolling resistance (KWsec/m), B: rotating term (KWsec^2/m^2), C: aerodynamic drag term (KWsec^3/m^3), g: acceleration due to gravity (9.81 m/s^2) and θ_t: road grade (degrees).

After finding the emissions per second corresponding to the OpModes from the emission factor table of the corresponding vehicle type, the emissions from the vehicle are calculated through the aggregation process. When it is necessary to aggregate the emissions by road section on the traffic network, the location information per second can be utilized to match the link ID suitable for the node ID and link ID of the traffic network, and the emissions for each link can be calculated.

3.2. Extraction of Representative OpMode Distribution and Estimation of Micro-Level Emission Factors

3.2.1. Concept

Emissions from individual vehicles can be estimated with vehicle trajectory data collected from GPS devices by the method described in Section 3.1. The problem is that the total amount of emissions cannot be calculated because it is not possible to collect the driving trajectory data from all vehicles in the target area. One way to consider is to make the extracted vehicle trajectories from the results of the microscopic traffic simulation network model, which can replace the actual vehicle trajectories. However, it has several disadvantages. First, it is not easy to calibrate a traffic simulation network model to make it similar to the actual traffic network. Second, if the target area is changed, the method can be applied only after establishing a new simulation network for that area. Furthermore, even if this method enables all the vehicle trajectories of the entire network to be acquired, calculating emissions from all vehicles in a large network using the micro-level emission estimation method would require a lot of time.

Another alternative is to find representative vehicle trajectories on the target road section at the analysis time interval for each vehicle type (passenger cars, trucks, buses), calculate emissions from those vehicles' trajectory data, and use the calculation results to estimate emissions from all vehicles in the road section at the analysis time interval. Because the traffic volume for each vehicle type in the analysis time interval of the target road can be obtained by using ITS equipment or the traffic volume estimation model, the total emissions can be calculated by multiplying the traffic volume by the emissions from the representative vehicle trajectories. This method has the advantage of higher efficiency in calculating regional vehicle emissions because the calculated emissions based on the representative vehicle trajectories from each road section can be used as emission factors (g/veh) for each road section. On that basis, in this study, this method has been applied to extract the trajectory of a vehicle, which has representative driving characteristics, from the vehicle trajectory data of some vehicles that passed the corresponding road section at the corresponding time interval.

Several studies aimed at developing representative vehicle trajectories have been established. For example, a representative link driving cycle by vehicle type, road type, and LOS has been developed for MOVES. However, it is not appropriate to apply their driving cycles to estimate regional vehicle emissions. The first reason is that the concept of the measurement link of the MOVES driving cycle is different from the link in the transportation network of the node link system. The former is closer to the travel route concept, while the latter refers to a road section whose length can vary and is shorter than a travel route. The second reason is that the concept of driving cycle is different between the MOVES driving cycles and the extracted representative link vehicle trajectories in this study. The MOVES driving cycles are developed to represent vehicle driving characteristics of various road types. On the other hand, this study extracts the representative vehicle trajectories to reflect the various driving situations that can appear in each link based on the driving data collected from each link. In other words, the driving characteristics of the same vehicle type that runs on the same road at the same time interval should have many similarities, as well as certain differences. For example, in the case of an arterial road, there will be a difference in vehicle trajectories between a vehicle experiencing traffic delay due to signals and one that is not. Thus, this study intends to classify vehicle trajectories into several similar groups and use the center of each group as a representative vehicle trajectory.

Figure 1 is a diagram showing the procedure for estimating the micro-level link emission factors required to calculate the emissions at the network level by using the collected vehicle trajectory data. The following section explains each process in detail.

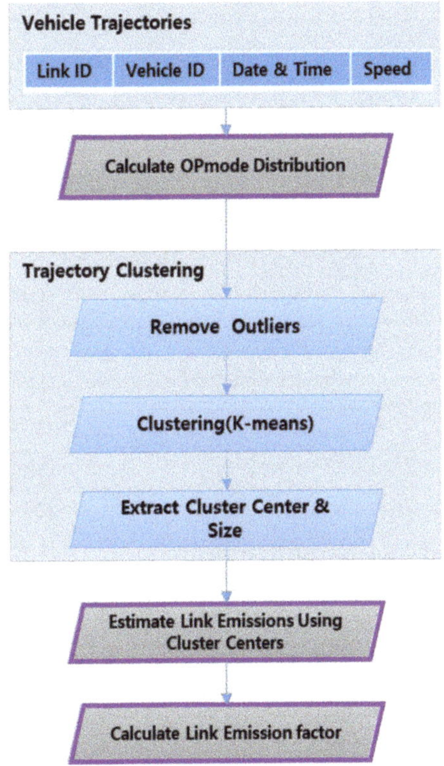

Figure 1. Process of Developing Link Emission Factors.

3.2.2. Calculate OpMode Distribution of Vehicle Trajectory

First, the vehicle trajectories of the same vehicle type on the same road section at the same time interval are collected. The vehicle trajectory data includes link ID, vehicle ID, recording time, and speed at each time point (in sec). Second, the VSP per second is calculated by the method described in Section 3.1.2 for each vehicle trajectory, and OpModes per second are classified for each vehicle trajectory. Then, the frequency for each of the 23 OpModes is calculated, and the ratios of each OpMode are calculated and stored. Each vehicle trajectory has 23 new variables, which are the ratios of each OpMode.

3.2.3. Trajectory Clustering

Before performing cluster analysis, outliers are removed from the vehicle trajectory dataset to filter out unusual driving cases that have not passed through the entire link or have stopped for a long time. Cluster analysis is performed to assemble the data into several similar vehicle trajectory groups. The k-means method is used to find the optimal number clusters and the center of each cluster. As a variable for each cluster analysis, the ratio by OpMode, which can describe the characteristics of the vehicle trajectory, is used. As in previous studies, aggregated characteristics, such as average speed and maximum acceleration, could be used as variables for clustering, while the ratio for each OpMode was selected as a cluster analysis variable. The main purpose is to utilize the value of the center of each group, which is the OpMode distribution, for micro-level emission calculation. Another purpose is to use the same scaled cluster variables. Using the values of the cluster center, the emissions of the cluster center can be calculated. The weighted average of emissions of all vehicles is obtained by applying the cluster size as a weight, which means the emissions per vehicle on the corresponding road section. It can be used as a micro-level link emission factor.

3.2.4. Calculate Link Emission Factor

After cluster analysis, the OpMode distribution and cluster size of each cluster center are extracted, which are used to calculate the emissions of all clusters by Equation (2):

$$\text{Emissions}_{pol} = \sum_{1}^{k}\left(\left(\sum^{opmode} R_{opmode} * EF_{opmode,\, pol}\right)_{k} * size_{k} * meanTT\right), \qquad (2)$$

where $\left(\sum^{opmode} R_{opmode} * EF_{opmode,\, pol}\right)_{k}$: represents the emissions of cluster k, R_{opmode} is the ratio of each OpMode, $EF_{opmode,\, pol}$ the OpMode emission factors by pollutants (pol, in g/sec), k is the cluster number, $size_{k}$ is the cluster size and $meanTT$ is the average driving time of vehicle trajectories

Finally, as shown in Equation (3), by dividing the total emissions by the total number of vehicles used in the cluster analysis, the micro-level link emission factor of the vehicle type is calculated by the following:

$$micLEF_{pol} = \text{emissions}_{pol}\,/\,\text{number of vehicles} \qquad (3)$$

where $micLEF_{pol}$: micro-level emission factor (g/veh).

According to the collected vehicle trajectory data, the data will be spatio-temporally expanded and applied to the regional and national levels. A database of micro-level link emission factors for a nationwide traffic network can be established. The procedure for estimating the micro-level link emission factors should be performed by vehicle type and analysis time interval on each road section. When a vehicle type-specific traffic volume database is provided, the emissions from all the vehicles on the traffic network will be easily calculated by applying the database of micro-level link emission factors for a nationwide traffic network.

4. Case Study

In the case analysis, emissions were estimated based on the method described in Section 3 by using the navigation data collected from actual roads, and it was investigated whether such a method produces accurate emission calculation results as intended. As the target roads, as shown in Figure 2, one section of the highway (about 1 km long) and one section of the arterial road (about 600 m long) in Bucheon, Gyeonggi-do in the Republic of Korea are selected to inspect whether the method reflects the driving characteristics of uninterrupted flow and interrupted flow, respectively. The target road sections are shown in Figure 2, and their attributes are described below.

Figure 2. Study Area. (Map source: Google Maps).

- Selected freeway link

  ○ Road name: Seoul Jeil Sunwhan Expressway
  ○ Number of lanes: Four per direction
  ○ Speed limit: 100 km/h

- Selected arterial link

  ○ Road name: Sosaro
  ○ Number of lanes: two per direction
  ○ Speed limit: 60 km/h

4.1. Data Collection

The navigation data was acquired in December 2017, and among the data on 13 December (Wednesday), the driving data on the morning peak hours (07:00–09:00), non-peak hours (13:00–15:00), and afternoon peak hours (17:00–19:00) for the northbound and southbound parts of each road section was extracted. Table 1 summarizes the number of vehicle trajectories collected for each analysis unit, showing that the vehicle trajectories corresponding to approximately 2–7% of the traffic volume were collected. After extracting the vehicle position in seconds from the navigation data, the data was organized into vehicle-specific driving trajectory data to calculate the speed per second and acceleration per second. Because most of the navigation data was provided from passenger cars, the analysis was conducted by considering the passenger car as the vehicle type.

4.2. Cluster Analysis Results

Cluster analysis was performed for each of the 12 groups listed in Table 1, and Table 2 summarizes the results. Among the collected vehicle trajectories, the data showing outliers in terms of the travel time was removed before the cluster analysis. Most of the removed vehicle trajectories were from vehicles that did not pass through all sections of the corresponding road. Table 2 shows the criteria for outlier removal and the number of selected vehicle trajectories. The cluster analysis results reveal that the best number of clusters for highway data is 7 to 8 and the ratio of goodness of fit is about 0.6 to 0.8. The results also show that the best number of clusters for the arterial road data is 6 to 8 and the ratio of goodness of fit is about 0.8 to 0.9.

Table 1. Acquired vehicle trajectories.

			Total Number of Trajectories	Traffic Volume [4] (veh/hour)	Collecting Ratio
Freeway Link	North bound	AM peak [1]	288	7128.5	2.0%
		Off peak [2]	430	7284.5	3.0%
		PM peak [3]	382	7527	2.5%
	South bound	AM peak	364	4487	4.1%
		Off peak	608	4062	7.5%
		PM peak	508	4022.5	6.3%
Arterial Links	North bound	AM peak	51	1375.5	1.9%
		Off peak	83	1440.5	2.9%
		PM peak	98	1635.5	3.0%
	South bound	AM peak	50	1709	1.5%
		Off peak	81	1268.5	3.2%
		PM peak	98	1289	3.8%

[1] 07:00–09:00, [2] 13:00–15:00, [3] 17:00–19:00, [4] Estimated traffic volume provided by the Korean Traffic Database.

Table 2. Summary of trajectory clustering.

	Freeway					
	Northbound			Southbound		
Time Period	AMpeak	offpeak	PMpeak	AMpeak	offpeak	PMpeak
Total trajectories	288	430	382	364	608	508
ATT, Average travel time (s)	111.84	138.16	90.86	234.8	163.77	176.55
Outlier upper limit (s)	−82.5	−70.75	−80.5	−360.5	−88	−178
Outlier lower limit (s)	217.5	207.25	223.5	819.5	360	526
Selected trajectories	74	134	92	95	110	125
ATT (s) of selected trajectories	103.27	99.26	107.16	93.18	90.8	100.46
Best number of Clusters	7	7	8	8	7	8
Clustering Fitness	0.7944	0.7068	0.6906	0.7131	0.6563	0.6269

	Arterial					
	Northbound			Southbound		
Time period	AMpeak	offpeak	PMpeak	AMpeak	offpeak	PMpeak
Total trajectories	51	83	98	50	81	98
ATT, Average travel time (s)	279.1	286.19	318.23	280.9	271.28	306.7
Outlier upper limit (s)	−439.75	−45.88	−60.75	−449	−81	−67
Outlier lower limit (s)	998.75	388.38	621.25	983	378	597
Selected trajectories	22	23	21	19	20	21
ATT (s) of selected trajectories	515.05	193.7	240.29	525.21	185.8	243.43
Best number of Clusters	6	6	8	7	6	8
Clustering Fitness	0.9284	0.8407	0.8693	0.9502	0.8001	0.9033

4.3. Emission Estimation Results

The emissions of air pollutants (CO, NO_x, PM_{10}, $PM_{2.5}$, and CO_2) were estimated based on the method described in Section 3. The estimated emissions were compared with the results obtained by estimating emissions using each individual vehicle trajectory through the micro-level emission estimation method of MOVES. As shown in Table 3, the difference in the emissions calculated by the two methods is insignificant, at 1–4% for the highway and 1–6% for the arterial road. The results prove that the proposed method has acceptable accuracy in estimating emissions.

Table 3. Comparison of link emission estimation results.

	AMpeak			offpeak			PMpeak		
Freeway Northbound									
	A [1]	B [2]	R [3]	A	B	R	A	B	R
CO (g)	94.6	97.55	1.03	156.98	158.23	1.01	113.87	116.67	1.02
NO_x (g)	4.46	4.624	1.04	8.128	8.248	1.01	5.868	6.091	1.04
$PM_{2.5}$ (g)	0.0227	0.0236	1.04	0.0366	0.0369	1.01	0.0254	0.0261	1.03
PM_{10} (g)	0.0254	0.0265	1.04	0.0414	0.0417	1.01	0.0282	0.029	1.03
CO_2 (g)	16,417.48	16,636.82	1.01	29,423.6	29,484.94	1.00	21,554.42	21,865.07	1.01
Freeway Southbound									
	A	B	R	A	B	R	A	B	R
CO (g)	124.12	125.22	1.01	128.17	129.48	1.01	149.25	150.85	1.01
NO_x (g)	5.994	6.047	1.01	6.943	7.05	1.02	7.957	8.107	1.02
$PM_{2.5}$ (g)	0.0309	0.0313	1.01	0.0299	0.0302	1.01	0.0335	0.0339	1.01
PM_{10} (g)	0.0348	0.0352	1.01	0.0334	0.0337	1.01	0.0373	0.0378	1.01
CO_2 (g)	20,139.64	20,127.88	1.00	23,156.19	23,252.67	1.00	28,295.29	28,363.63	1.00
Arterial Northbound									
	A	B	R	A	B	R	A	B	R
CO (g)	53.48	56.78	1.06	27.45	28.52	1.04	29.06	30.12	1.04
NO_x (g)	1.931	2.043	1.06	1.012	1.061	1.05	1.051	1.089	1.04
$PM_{2.5}$ (g)	0.0153	0.0158	1.03	0.0067	0.0069	1.03	0.0075	0.0077	1.03
PM_{10} (g)	0.0223	0.0228	1.02	0.0089	0.0091	1.02	0.01	0.0102	1.02
CO_2 (g)	13,893.62	14,425.56	1.04	6248.72	6437.99	1.03	6790.39	7006.47	1.03
Arterial Southbound									
	A	B	R	A	B	R	A	B	R
CO (g)	47.46	48.7	1.03	24.04	24.32	1.01	29.91	30.25	1.01
NO_x (g)	1.719	1.786	1.04	0.876	0.896	1.02	1.086	1.098	1.01
$PM_{2.5}$ (g)	0.0137	0.0138	1.01	0.0058	0.0059	1.02	0.0077	0.0077	1.00
PM_{10} (g)	0.02	0.02	1.00	0.0076	0.0076	1.00	0.0104	0.0104	1.00
CO_2 (g)	12,413	12,601.47	1.02	5398.06	5422.57	1.00	7041.24	7069.78	1.00

[1] Estimating emissions using each individual vehicle trajectory through the micro-level emission estimation method of MOVES, [2] Estimating emissions using emissions of cluster centers through the proposed method, [3] B/A.

4.4. Micro-Level Link Emission Factors

The micro-level link emission factors are calculated by emissions from vehicles in all clusters, which are B columns in Table 3 by the number of vehicles (the number of vehicle trajectories), and summarized in Table 4. These values are the micro-level emission factors of the passenger car at the analysis time interval on the analysis link for each pollutant. As shown in Figure 3, plotting these values through bar graphs can determine whether the estimated micro-level link emission factors can appropriately reflect the emission characteristics of the link. The emission factors show the highest trend in the morning peak hours and the lowest in the non-peak hours, indicating the change in emissions according to time periods of the link. Even though the driving length of the highway (about 1 km) is longer than that of the arterial road (about 600 m), the emissions are lower on the highway, which indicates that the emission factors appropriately reflect the driving characteristics by the road characteristics (uninterrupted flow and interrupted flow) of the link.

Table 4. Estimated link emission factors (g/veh).

	Freeway					
	Northbound			Southbound		
Time Period	AMpeak	offpeak	PMpeak	AMpeak	offpeak	PMpeak
CO	1.32	1.18	1.27	1.32	1.18	1.21
NO_x	0.0625	0.0616	0.0662	0.0637	0.0641	0.0649
$PM_{2.5}$	0.000319	0.000275	0.000284	0.000329	0.000275	0.000271
PM_{10}	0.000358	0.000311	0.000315	0.000371	0.000306	0.000302
CO_2	224.82	220.04	237.66	211.87	211.39	226.91
	Arterial					
	Northbound			Southbound		
Time Period	AMpeak	offpeak	PMpeak	AMpeak	offpeak	PMpeak
CO	2.58	1.24	1.43	2.56	1.22	1.44
NO_x	0.0929	0.0461	0.0519	0.0940	0.0448	0.0523
$PM_{2.5}$	0.000718	0.000300	0.000367	0.000726	0.000295	0.000367
PM_{10}	0.001036	0.000396	0.000486	0.001053	0.000380	0.000495
CO_2	655.71	279.91	333.64	663.24	271.13	336.66

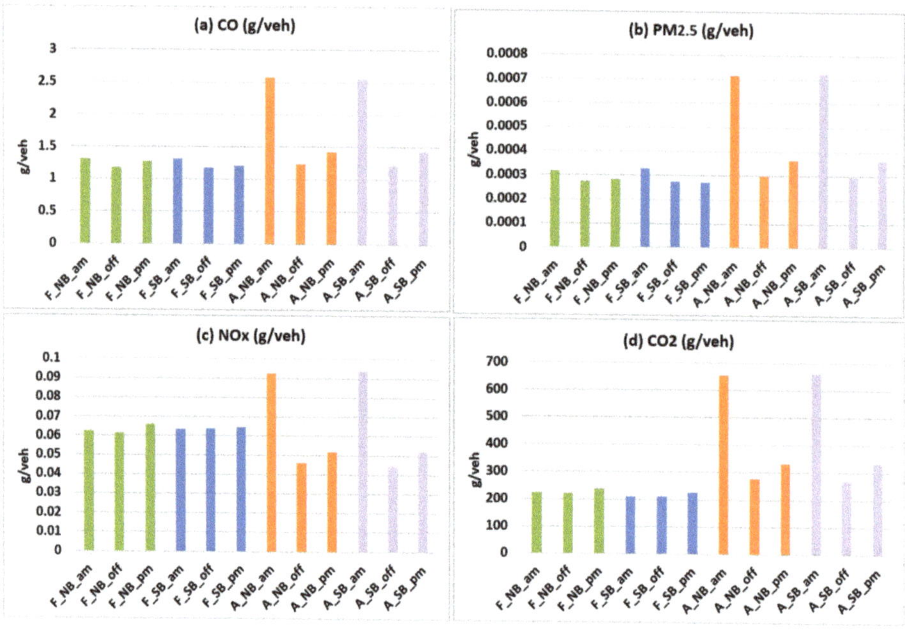

Figure 3. Estimated Link Emission Factors by Pollutant Type.

5. Effects on Clustering and Micro-Level Link Emission Factors by Using Accumulated Vehicle Trajectories

This study has confirmed that the proposed method can be applied to estimate the total emissions from vehicles traveling on a road section with the actual vehicle trajectory data through the case study. Micro-level link emission factors for links were derived through the proposed emission estimation process using vehicle trajectories collected from one day. This method may be applied by using accumulated vehicle trajectory data collected over several days to increase the micro-level link emission factors' representativeness. Use of an accumulated dataset can offset the limitations of vehicle trajectory data, which have a low rate of acquisition, and enable the analysis time period to be divided into shorter periods to increase detail.

To analyze the effect of using accumulated vehicle trajectory data, the vehicle trajectory data of the freeway northbound link selected for the case study was additionally acquired. The trajectory data of the vehicles traveling through the road section during the 2-h morning peak on weekdays (Tuesday, Wednesday, and Thursday) in December 2017 was used. A total of 12 datasets were made by increasing the number of data collection days from one day to 12 days. For example, the first dataset included one day of data, the second dataset included an additional day of data, and so on. The proposed method was applied to each of those datasets.

In this analysis, changing patterns in cluster analysis results and estimated micro-level emission factors were investigated by accumulating daily data. Table 5 shows the number of remaining vehicle trajectories after removing outliers among the vehicle trajectories collected on each date. Table 6 summarizes the cluster analysis results subject to daily accumulation of the data, which shows that the number of vehicle trajectories used for cluster analysis increases as the daily data is accumulated, reaching 1029 after the 12th day of accumulation. The average travel times of the vehicle trajectories used for cluster analysis show a pattern that converges to about 101 s. The optimal number of clusters is 7, and the ratio of goodness of fit is 0.7 or more.

Table 5. Vehicle trajectory selection results.

Days	12/5	12/6	12/7	12/12	12/13	12/14
Total trajectories	301	253	264	288	288	282
ATT, Average travel time (s)	102.41	104.87	125.31	134.15	111.84	106.4
Outlier upper limit (s)	−66.25	−92.12	−66.5	−71.5	−82.5	−59
Outlier lower limit (s)	196.25	247.12	217.5	220.5	217.5	205
Selected trajectories	80	67	81	93	74	91
ATT (s) of selected trajectories	92.85	112.73	103.69	100.72	103.27	98.21
Days	**12/19**	**12/20**	**12/21**	**12/26**	**12/27**	**12/28**
Total trajectories	283	303	246	261	300	314
ATT, Average travel time (s)	114.8	93.67	87.94	98.64	95.41	92.2
Outlier upper limit (s)	−59.88	−59.88	−74.5	−64	−71	−84.25
Outlier lower limit (s)	205.38	201.38	217.5	192	209	225.75
Selected trajectories	101	92	72	76	103	99
ATT (s) of selected trajectories	102.94	97.85	104.21	93.21	103.65	104.32

Table 6. Summary of trajectory clustering by increasing number of data collection days.

Days	1 day	2 days	3 days	4 days	5 days	6 days
Selected trajectories	80	147	228	321	395	486
ATT (s) of selected trajectories	92.85	101.91	102.54	102.02	102.25	101.49
Best number of clusters	7	7	7	7	7	7
Clustering Fitness	0.7195	0.7723	0.7474	0.7563	0.754	0.7379
Days	**7 days**	**8 days**	**9 days**	**10 days**	**11 days**	**12 days**
Selected trajectories	587	679	751	827	930	1029
ATT (s) of selected trajectories	101.74	101.22	101.5	100.74	101.06	101.38
Best number of Clusters	7	7	7	7	7	7
Clustering Fitness	0.7345	0.7343	0.729	0.7244	0.7254	0.7228

Figure 4 is a diagram comparing the OpMode distributions of seven cluster centers derived from each dataset. It can be observed that the shapes of the OpMode distribution of the seven cluster centers are similar after Day 4. This graph offers useful information for determining how many days are required to accumulate vehicle trajectories for estimating micro-level link emission factors.

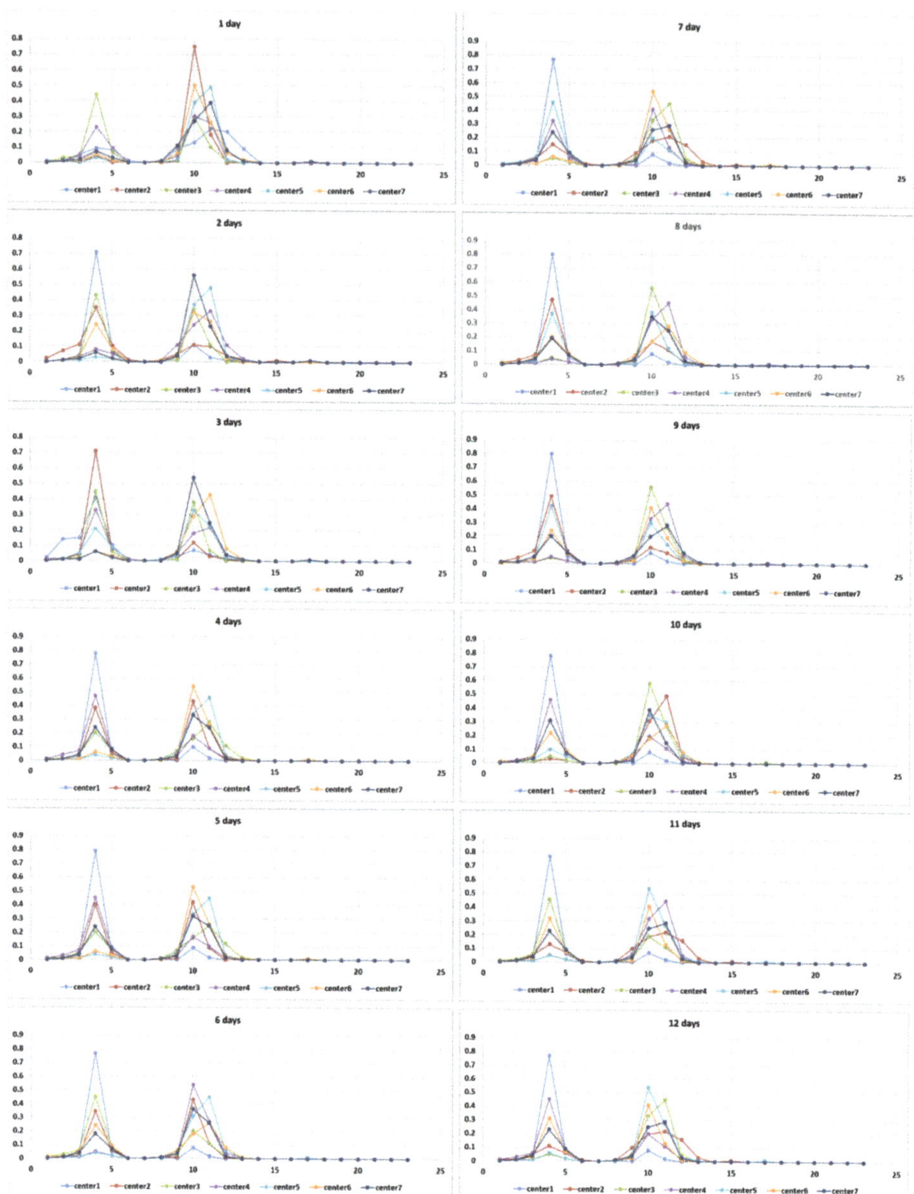

Figure 4. OpMode Distributions of Cluster Centers.

Table 7 summarizes the micro-level emission factors for each pollutant estimated by adding the number of data collection days. The bar graphs plotted in Figure 5 show that the micro-level emission factors tend to converge as the number of days of data accumulation increases. It means that the estimated micro-level link emission factors' representativeness increases as the data is accumulated.

Table 7. Estimated link emission factors (g/veh) by adding number of data collection days.

Days	1 day	2 days	3 days	4 days	5 days	6 days
CO	1.398625	1.360816	1.32943	1.321153	1.320658	1.326461
NO_x	0.0661	0.064524	0.062522	0.062414	0.06243	0.062914
$PM_{2.5}$	0.000356	0.000335	0.000325	0.000323	0.000322	0.000324
PM_{10}	0.000403	0.000379	0.000368	0.000366	0.000364	0.000366
CO_2	213.4434	224.6081	223.2107	222.4063	222.8595	222.4115

Days	7 days	8 days	9 days	10 days	11 days	12 days
CO	1.323799	1.326966	1.322703	1.33162	1.326516	1.317891
NO_x	0.062833	0.062929	0.062848	0.063204	0.063115	0.062888
$PM_{2.5}$	0.000323	0.000324	0.000323	0.000326	0.000324	0.000321
PM_{10}	0.000365	0.000367	0.000365	0.000368	0.000366	0.000363
CO_2	222.6467	222.0059	222.3667	221.6501	222.0375	222.2886

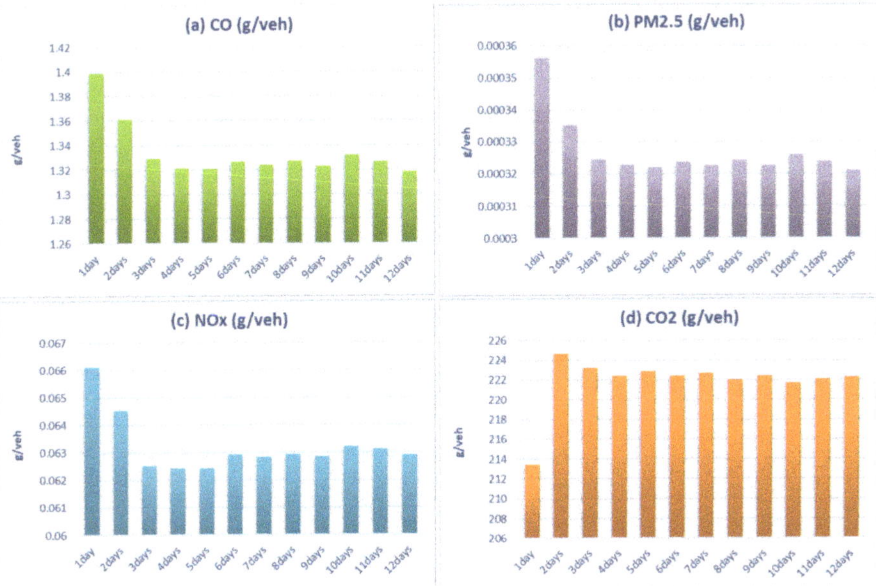

Figure 5. Changes in Estimated Link Emission Factors by Increasing Number of Data Collection Days.

6. Discussion

The applicability of the developed methodology was examined by using the navigation data collected according to each of the three (morning peak hours, non-peak hours, afternoon peak hours) analysis time intervals on one highway section (1 km) and one arterial road section (600 m). The results of the analysis confirmed that the error rate showed a difference in the range of 1–4% for the highway section and 1–6% for the arterial road section when the link emissions were calculated using the proposed method. Moreover, the results indicated that the estimated micro-level link emission factors reflect the driving characteristics of the link according to the traffic conditions for each time period and the driving characteristics for each road characteristic (uninterrupted flow and interrupted flow). Additionally, the analysis results through the same method while accumulating the vehicle trajectory data showed that clustering analysis produced similar cluster centers after 4-day data accumulation and link emission factors were converged to certain values representing the vehicle travel characteristics on the corresponding road at the corresponding time period by adding the number of data collection days. The number of days to reach convergence on the target road in this study was about four, while it is expected to increase in road sections having fewer collected vehicle trajectories. These results mean that the proposed method can offset the limitations of vehicle trajectory data with a small number of samples. Furthermore, this approach enables the analysis time period to be shorter and thus the micro-level emission calculation of the entire network to be analyzed in more detail.

In this study, because navigation data was used, the results were limited to the case of passenger cars. Therefore, it is necessary to acquire the vehicle trajectories of a commercial vehicle, such as a truck or bus, from DTG data to check whether the applicability is the same. In addition, considering that the link length of the traffic network varies, it is also necessary to vary the collected link length to check whether there are any points to be supplemented in the analysis method. Moreover, it is also required to confirm the basis for judgment as to how many days for data accumulation and analysis would enable representative link micro-level emission factors to be developed.

7. Conclusions

Several problems must be addressed to apply the micro-level emission estimation method at the regional or national level by using the vehicle trajectory data collected through GPS data. The biggest problem is that it is not possible to collect the vehicle trajectory data of all vehicles running on the traffic network. The second problem is related to the task of extracting necessary data from the collected vehicle trajectory, which requires a considerable amount of data processing and operation time in calculating the micro-level emissions of individual vehicles and aggregating results by road section.

This study proposed a countermeasure to solve these problems. In this study, a micro-level emission estimation method using the massive vehicle trajectory data collected from vehicle navigation, DTG, and mobile devices was developed, which can be applicable at the regional or national level. The vehicle trajectories from collected GPS data were classified as link ID and time period to estimate the emissions and emission factors for each link at a specific time period. Vehicle trajectories for a link at a time period were divided into several groups through cluster analysis, in which the ratios of each OpMode used in MOVES were used as cluster variables for clustering similar vehicle trajectories. The choice of cluster variables is the biggest difference from the other methods for clustering vehicle trajectories. The derived values of each cluster center from clustering analysis, the OpMode distribution, can be used for calculating micro-level emissions. The center of each cluster denotes the representative vehicle trajectory for each cluster. The emissions of the cluster center can be calculated easily by using the values of the cluster center. The weighted averages of emissions of all vehicles are obtained by applying the cluster size as a weight, which represents the emissions per vehicle on the corresponding road section. They can be used as micro-level link emission factors to estimate emissions of regional- or national-level traffic networks. When vehicle type-specific traffic volume is provided, the emissions from all the vehicles on the traffic network will be easily calculated by multiplying by the micro-level link emission factors. This is the main purpose of developing the proposed method.

The proposed method is not free from computational difficulty because the operating distribution of each vehicle trajectory must be calculated to estimate link-based micro-level emission factors. Moreover, more data must be collected and analyzed in order to increase the representativeness. This requires more storage space and computing power. Fortunately, not only can the calculation procedures of the proposed method be automated but also high-performance machines can be utilized for the calculation. Thus, it is expected that the issues of storage space and computing power related to the proposed method can be addressed.

The confirmation procedure explained in the previous section is still required. However, if the proposed method were automated to accumulate data, such as navigation data, DTG data, and mobile data, for each traffic network link and update the link-based micro-level emission factors, only having traffic volume by vehicle type at the analysis time period would enable local or nationwide micro-level emission estimation to be performed efficiently.

Electronics **2020**, *9*, 2151

Author Contributions: Conceptualization, H.H. and G.L.; Data curation, J.H.K. and H.S.; Formal analysis, H.H. and H.S.; Writing—original draft, H.H., J.H.K., and G.L.; supervision, G.L. All authors have read and agreed to the published version of the manuscript.

Funding: This research was supported by a grant(20TLRP-B148676-03) from Transportation & Logistics Research Program (TLRP) Program funded by Ministry of Land, Infrastructure and Transport of Korean government and also supported by supported by the National Research Foundation of Korea grant funded by the Republic of Korea government (MSIT) (No. 2018R1C1B6006330).

Conflicts of Interest: The authors declare no conflict of interest.

Appendix A

Table A1. Calculated Emission Rates (g/s) by OpMode for a Passenger Car in Gyeonggi-Do, Korea.

Operating Mode	Operating Mode Description		CO	NO$_X$	PM$_{2.5}$	PM$_{10}$	CO$_2$
0	Braking		0.002161	0.000115	0.000002	0.000002	0.884684
1	Idling		0.000954	0.000072	0.000001	0.000002	0.816689
11	VSP < 0		0.005995	0.000169	0.000001	0.000002	1.307695
12	0 ≤ VSP < 3		0.009848	0.000298	0.000002	0.000002	1.821388
13	3 ≤ VSP < 6	1 ≤ Speed < 25	0.010114	0.000661	0.000002	0.000002	2.537854
14	6 ≤ VSP < 9		0.0146	0.001173	0.000002	0.000002	3.206731
15	9 ≤ VSP < 12		0.020065	0.001875	0.000002	0.000003	3.832276
16	12 ≤ VSP		0.031874	0.0034	0.000007	0.000008	4.674586
21	VSP < 0		0.00833	0.000373	0.000003	0.000003	1.79065
22	0 ≤ VSP < 3		0.010655	0.000526	0.000003	0.000004	2.048499
23	3 ≤ VSP < 6		0.013619	0.00079	0.000003	0.000003	2.500449
24	6 ≤ VSP < 9		0.020059	0.001339	0.000003	0.000003	3.203335
25	9 ≤ VSP < 12	25 ≤ Speed < 50	0.022456	0.00187	0.000004	0.000004	4.285691
27	12 ≤ VSP < 18		0.034629	0.003022	0.000006	0.000006	5.646284
28	18 ≤ VSP < 24		0.072006	0.005967	0.000018	0.000019	7.557838
29	24 ≤ VSP < 30		0.152696	0.00988	0.000087	0.000094	10.39979
30	30 ≤ VSP		0.523226	0.012622	0.000166	0.00018	13.05113
33	VSP < 6		0.006355	0.000653	0.000004	0.000004	2.564789
35	6 ≤ VSP < 12		0.011256	0.00182	0.000005	0.000006	4.115463
37	12 ≤ VSP < 18	50 ≤ Speed	0.016235	0.002606	0.000005	0.000005	5.353319
38	18 ≤ VSP < 24		0.066225	0.005082	0.00001	0.000011	6.972214
39	24 ≤ VSP < 30		0.073859	0.007304	0.000023	0.000025	9.269841
40	30 ≤ VSP		0.206546	0.009329	0.000027	0.000029	11.81478

Appendix B

Table A2. Calculated Emission Rates (g/s) by OpMode for a Truck in Gyeonggi-Do, Korea.

Operating Mode	Operating Mode Description		CO	NO$_X$	PM$_{2.5}$	PM$_{10}$	CO$_2$
0	Braking		0.005157	0.001119	0.000011	0.000012	1.666371
1	Idling		0.004261	0.000736	0.000014	0.000015	1.235385
11	VSP < 0		0.010469	0.000899	0.000017	0.000018	1.838659
12	0 ≤ VSP < 3		0.017583	0.002737	0.000026	0.000028	2.656532
13	3 ≤ VSP < 6	1 ≤ Speed < 25	0.020172	0.00494	0.000114	0.000124	4.279017
14	6 ≤ VSP < 9		0.028832	0.007317	0.000177	0.000193	5.839649
15	9 ≤ VSP < 12		0.038281	0.009744	0.000279	0.000304	7.304234
16	12 ≤ VSP		0.055333	0.014565	0.000318	0.000346	9.664224

Table A2. *Cont.*

Operating Mode	Operating Mode Description		CO	NO$_X$	PM$_{2.5}$	PM$_{10}$	CO$_2$
21	VSP < 0		0.014596	0.001274	0.00001	0.000011	2.289468
22	0 ≤ VSP < 3		0.019214	0.00355	0.000028	0.000031	3.151216
23	3 ≤ VSP < 6		0.025321	0.005591	0.000084	0.000091	4.485333
24	6 ≤ VSP < 9		0.033635	0.008536	0.000191	0.000208	6.13888
25	9 ≤ VSP < 12	25 ≤ Speed < 50	0.039856	0.011277	0.00026	0.000282	7.872075
27	12 ≤ VSP < 18		0.055945	0.016247	0.000336	0.000365	10.90358
28	18 ≤ VSP < 24		0.099616	0.025651	0.000481	0.000523	14.60375
29	24 ≤ VSP < 30		0.218662	0.038046	0.000673	0.000732	19.44426
30	30 ≤ VSP		0.785452	0.047597	0.000983	0.001068	22.31806
33	VSP < 6		0.014684	0.003405	0.000055	0.00006	4.14084
35	6 ≤ VSP < 12		0.026154	0.010734	0.000166	0.000181	7.277741
37	12 ≤ VSP < 18	50 ≤ Speed	0.036623	0.01593	0.000212	0.000231	10.34293
38	18 ≤ VSP < 24		0.092717	0.025211	0.000265	0.000288	13.41121
39	24 ≤ VSP < 30		0.119214	0.034967	0.000353	0.000383	17.01697
40	30 ≤ VSP		0.31603	0.042564	0.000407	0.000442	21.822

Appendix C

Table A3. Calculated Emission Rates (g/s) by OpMode for a Bus in Gyeonggi-Do, Korea.

Operating Mode	Operating Mode Description		CO	NO$_X$	PM$_{2.5}$	PM$_{10}$	CO$_2$
0	Braking		0.007783	0.010314	0.000114	0.000123	3.681011
1	Idling		0.00391	0.006301	0.000155	0.000168	2.005807
11	VSP < 0		0.012418	0.00613	0.000174	0.000189	2.707361
12	0 ≤ VSP < 3		0.020119	0.023276	0.000245	0.000266	7.356385
13	3 ≤ VSP < 6	1 ≤ Speed < 25	0.053127	0.043837	0.001007	0.001094	14.87026
14	6 ≤ VSP < 9		0.081189	0.062972	0.00167	0.001815	22.47855
15	9 ≤ VSP < 12		0.110298	0.077616	0.002593	0.002819	29.30343
16	12 ≤ VSP		0.152142	0.105329	0.002968	0.003226	40.76052
21	VSP < 0		0.015794	0.004861	0.000075	0.000082	2.545149
22	0 ≤ VSP < 3		0.026166	0.027358	0.000221	0.00024	9.294441
23	3 ≤ VSP < 6		0.082666	0.046367	0.000703	0.000764	16.8997
24	6 ≤ VSP < 9		0.092526	0.066794	0.001734	0.001885	23.33269
25	9 ≤ VSP < 12	25 ≤ Speed < 50	0.124529	0.084568	0.002144	0.00233	30.56697
27	12 ≤ VSP < 18		0.16993	0.115507	0.002604	0.002831	42.55378
28	18 ≤ VSP < 24		0.230723	0.148917	0.003083	0.003351	59.52028
29	24 ≤ VSP < 30		0.304292	0.191851	0.003466	0.003767	76.57897
30	30 ≤ VSP		0.415553	0.234549	0.004119	0.004477	93.48057
33	VSP < 6		0.033908	0.028366	0.000531	0.000577	9.532979
35	6 ≤ VSP < 12		0.071659	0.077374	0.001637	0.00178	26.14048
37	12 ≤ VSP < 18	50 ≤ Speed	0.110222	0.123471	0.001924	0.002091	41.36648
38	18 ≤ VSP < 24		0.148665	0.163963	0.002131	0.002316	57.82597
39	24 ≤ VSP < 30		0.191404	0.211028	0.002395	0.002604	74.32924
40	30 ≤ VSP		0.2482	0.257894	0.0025	0.002718	90.92918

References

1. Lee, G.; You, S.; Ritchie, S.G.; Saphores, J.-D.; Sangkapichai, M.; Jayakrishnan, R. Environmental Impacts of a Major Freight Corridor: A Study of I-710 in California. *Transp. Res. Rec.* **2009**, *2123*, 119–128. [CrossRef]
2. EMFAC. Available online: https://arb.ca.gov/emfac/ (accessed on 14 November 2020).
3. National Institute of Environmental Research National Air Pollutant Emission Calculation Method Manual III. Available online: https://www.nier.go.kr/NIER/cop/bbs/selectNoLoginBoardList.do (accessed on 14 November 2020).
4. COPERT4. Road Transport Emissions Model—European Environment Agency. Available online: https://www.eea.europa.eu/themes/air/links/guidance-and-tools/copert4-road-transport-emissions-model (accessed on 14 November 2020).
5. Ahn, K.; Rakha, H.; Trani, A.; Van Aerde, M. Estimating Vehicle Fuel Consumption and Emissions based on Instantaneous Speed and Acceleration Levels. *J. Transp. Eng.* **2002**, *128*, 182–190. [CrossRef]
6. Barth, M.; Malcolm, C.; Younglove, T.; Hill, N. Recent Validation Efforts for a Comprehensive Modal Emissions Model. *Transp. Res. Rec.* **2001**, *1750*, 13–23. [CrossRef]
7. Beckx, C.; Panis, L.I.; Janssens, D.; Wets, G. Applying activity-travel data for the assessment of vehicle exhaust emissions: Application of a GPS-enhanced data collection tool. *Transp. Res. Part Transp. Environ.* **2010**, *15*, 117–122. [CrossRef]
8. Barth, M.J.; Johnston, E.; Tadi, R.R. Using GPS Technology to Relate Macroscopic and Microscopic Traffic Parameters. *Transp. Res. Rec.* **1996**, *1520*, 89–96. [CrossRef]
9. Rakha, H.; Medina, A.; Sin, H.; Dion, F.; Van Aerde, M.; Jenq, J. Traffic Signal Coordination Across Jurisdictional Boundaries: Field Evaluation of Efficiency, Energy, Environmental, and Safety Impacts. *Transp. Res. Rec.* **2000**, *1727*, 42–51. [CrossRef]
10. Gately, C.K.; Hutyra, L.R.; Peterson, S.; Wing, I.S. Urban emissions hotspots: Quantifying vehicle congestion and air pollution using mobile phone GPS data. *Environ. Pollut.* **2017**, *229*, 496–504. [CrossRef]
11. Kan, Z.; Tang, L.; Kwan, M.-P.; Zhang, X. Estimating Vehicle Fuel Consumption and Emissions Using GPS Big Data. *Int. J. Environ. Res. Public Health* **2018**, *15*, 566. [CrossRef]
12. Sui, Y.; Zhang, H.; Song, X.; Shao, F.; Yu, X.; Shibasaki, R.; Sun, R.; Yuan, M.; Wang, C.; Li, S.; et al. GPS data in urban online ride-hailing: A comparative analysis on fuel consumption and emissions. *J. Clean. Prod.* **2019**, *227*, 495–505. [CrossRef]
13. US EPA Latest Version of MOtor Vehicle Emission Simulator (MOVES). Available online: https://www.epa.gov/moves/latest-version-motor-vehicle-emission-simulator-moves (accessed on 14 November 2020).
14. Lee, J.; Choi, J.-S.; Hu, H.; Yoon, T. A method for the estimation of greenhouse gas emissions based on road geometric design and its application to South Korea. *Int. J. Sustain. Transp.* **2019**, *13*, 65–80. [CrossRef]
15. Joo, S.; Oh, C.; Lee, S.; Lee, G. Assessing the impact of traffic crashes on near freeway air quality. *Transp. Res. Part Transp. Environ.* **2017**, *57*, 64–73. [CrossRef]
16. Liu, H.; Chen, X.; Wang, Y.; Han, S. Vehicle Emission and Near-Road Air Quality Modeling for Shanghai, China: Based on Global Positioning System Data from Taxis and Revised MOVES Emission Inventory. *Transp. Res. Rec.* **2013**, *2340*, 38–48. [CrossRef]
17. Tong, R.; Liu, J.; Wang, W.; Fang, Y. Health effects of PM2.5 emissions from on-road vehicles during weekdays and weekends in Beijing, China. *Atmos. Environ.* **2020**, *223*, 117258. [CrossRef]
18. Pu, Y.; Yang, C.; Liu, H.; Chen, Z.; Chen, A. Impact of license plate restriction policy on emission reduction in Hangzhou using a bottom-up approach. *Transp. Res. Part Transp. Environ.* **2015**, *34*, 281–292. [CrossRef]
19. Wu, Y.; Song, G.; Yu, L. Sensitive analysis of emission rates in MOVES for developing site-specific emission database. *Transp. Res. Part D* **2014**, *32*, 193–206. [CrossRef]
20. Perugu, H. Emission modelling of light-duty vehicles in India using the revamped VSP-based MOVES model: The case study of Hyderabad. *Transp. Res. Part D* **2019**, *68*, 150–163. [CrossRef]
21. Shan, X.; Chen, X.; Jia, W.; Ye, J. Evaluating Urban Bus Emission Characteristics Based on Localized MOVES Using Sparse GPS Data in Shanghai, China. *Sustainability* **2020**, *11*, 2936. [CrossRef]
22. Abou-Senna, H.; Radwan, E.; Westerlund, K.; Cooper, D. Using a traffic simulation model (VISSIM) with an emissions model (MOVES) to predict emissions from vehicles on a limited-access highway. *J. Air Waste Manag. Assoc.* **2013**, *63*, 819–831. [CrossRef]

23. Pouresmaeili, M.A.; Aghayan, I.; Taghizadeh, S.A. Development of Mashhad driving cycle for passenger car to model vehicle exhaust emissions calibrated using on-board measurements. *Sustain. Cities Soc.* **2018**, *36*, 12–20. [CrossRef]

24. Achour, H.; Olabi, A.G. Driving cycle developments and their impacts on energy consumption of transportation. *J. Clean. Prod.* **2016**, *112*, 1778–1788. [CrossRef]

25. Frey, H.C.; Liu, B. Development and Evaluation of Simplified Version of MOVES for Coupling with Traffic Simulation Model. In Proceedings of the Annual Meeting of Transportation Research Board, Washington, DC, USA, 13–17 January 2013.

26. Lee, G.; Joo, S.; Oh, C.; Choi, K. An evaluation framework for traffic calming measures in residential areas. *Transp. Res. Part Transp. Environ.* **2013**, *25*, 68–76. [CrossRef]

27. Fu, Z.; Hu, W.; Tan, T. Similarity Based Vehicle Trajectory Clustering and Anomaly Detection. *IEEE. Int. Conf. Image Process.* **2005**, 8845766. [CrossRef]

28. Song, H.; Wang, X.; Hua, C.; Wang, W.; Guan, Q.; Zhang, Z. Vehicle trajectory clustering based on 3D information via a coarse-to-fine strategy. *Soft Comput.* **2018**, *22*, 1433–1444. [CrossRef]

29. Fotouhi, A.; Montazeri-Gh, M. Tehran driving cycle development using the k-means clustering method. *Sci. Iran.* **2013**, *20*, 286–293. [CrossRef]

30. US EPA MOVES and Other Mobile Source Emissions Models. Available online: https://www.epa.gov/moves (accessed on 14 November 2020).

Publisher's Note: MDPI stays neutral with regard to jurisdictional claims in published maps and institutional affiliations.

 *electronics*

Article

For Preventative Automated Driving System (PADS): Traffic Accident Context Analysis Based on Deep Neural Networks

Minhee Kang [1], Jaein Song [2] and Keeyeon Hwang [3,*]

[1]   Department of Smartcity, Hongik University, Seoul 04066, Korea; speakingbee@hanmail.net
[2]   Department of Urban Planning, Hongik University, Seoul 04066, Korea; wodlsthd@nate.com
[3]   Department of Urban Design & Planning, Hongik University, Seoul 04066, Korea
*   Correspondence: keeyeonhwang@gmail.com; Tel.: +82-010-8654-7415

Received: 5 October 2020; Accepted: 30 October 2020; Published: 2 November 2020

Abstract: Automated Vehicles (AVs) are under development to reduce traffic accidents to a great extent. Therefore, safety will play a pivotal role to determine their social acceptability. Despite the fast development of AVs technologies, related accidents can occur even in an ideal environment. Therefore, measures to prevent traffic accidents in advance are essential. This study implemented a traffic accident context analysis based on the Deep Neural Network (DNNs) technique to design a Preventive Automated Driving System (PADS). The DNN-based analysis reveals that when a traffic accident occurs, the offender's injury can be predicted with 85% accuracy and the victim's case with 67%. In addition, to find out factors that decide the degree of injury to the offender and victim, a random forest analysis was implemented. The vehicle type and speed were identified as the most important factors to decide the degree of injury of the offender, while the importance for the victim is ordered by speed, time of day, vehicle type, and day of the week. The PADS proposed in this study is expected not only to contribute to improve the safety of AVs, but to prevent accidents in advance.

Keywords: preventive automated driving system; automated vehicle; traffic accidents; deep neural networks

1. Introduction

Amid an active discussion of the Fourth Industrial Revolution, Automated Vehicles (AVs) are expected to play an important role in leading the Fourth Industrial Revolution. AVs are defined as vehicles capable of navigating, controlling, and avoiding risk partly or totally without human assistance [1]. According to the Society of Automotive Engineering [2], AVs can be categorized into six levels, ranging from none auto-system (SAE level 0) to full auto-system (SAE level 5). Human intervention is minimized from SAE level 3 and driverless driving is possible at level 5. With such features as driving safety improvement, increase in convenience and mobility [3,4], AVs are highly evaluated as key future mobility of reducing traffic accidents. The benefits mentioned above will be accomplished when AVs fully take root. However, some researches have indicated that the public still expresses a low level of acceptance for AVs [5–7]. It is mainly attributed to AVs traffic accidents arisen during the test driving by Google, Uber, etc. In particular, a fatal pedestrian accident involving Uber has been at the forefront of ethical controversy over AVs. Neither of the types of traditional ethics (deontology, utilitarianism) fit in well to provide a proper answer to this accident, nor the trolley dilemma excuse is unsuitable [8–10]. In response, the AVs guidelines, including provisions of preventive design and safety, were issued in Germany and the United States [11,12]. Specifically, German AV ethics guidelines state that "Automated and connected technology should prevent accidents wherever this is practically possible" in its fifth clause. In addition, various studies emphasized that trust in AVs is the most important determinant to accept AVs for their mobility, and that the trust is decided by perceived safety risk, compatibility, and system quality [13–15].

To respond to the prevention of potential AVs related accidents, this study proposes Preventive Automated Driving System (PADS) of using Deep Neural Networks (DNNs)-based traffic accident context analysis. The study conducts experiments to identify key features affecting traffic accidents caused by unpredictable conditions such as black ice, sink-hall, centerline crossing, and so on.

The paper is organized as follows: Section 2 reviews studies on the effects of AVs and various traffic accident cases, and examines deep learning applications used in transportation research. Section 3 introduces methods to collect and process traffic accident data. In Section 4, we introduce how to build an optimal DNNs algorithm for forecasting the severity of accident injuries and extract factors causing accidents. Section 5 validates the important factors extracted in Section 4 by using a random forest-based machine-learning algorithm. Finally, Section 6 concludes the paper with a summary of empirical findings and derives future researches and implications related to preventive AVs.

2. Literature Review

2.1. AVs Introduction Impacts

It is important to estimate AV's impacts on traffic because it is inevitable that AVs will be mixed with human driving vehicles (HVs) in road traffic. With AV's Market Penetration Rate (MPR) growing, the positive effects can become bigger in congested conditions [16]. Most studies have been conducted quantitatively using a microscopic simulation software, VISSIM: A Leksandra Deluka Tibljaš et al. [17] designed a rotary interchange and evaluated safety in a mixed traffic condition. Yan Wang et al. [18] confirmed that AVs introduction not only kept Level of Service (LOS) higher, but also improved safety at signaled intersections. However, when MPR reached over 50%, a negative impact began to appear with a growing traffic delay [19]. Lee at al. [20] also identified that the maneuvering of AVs should be properly controlled by various traffic and road conditions because the driving behavior of HVs is affected by the aggressiveness of AVs. Some studies have mentioned that AVs contribute to decrease traffic accidents (collision) and delays [21]. Kolarova et al. [22] conducted an online survey for analyzing the potential changes in the Value of Travel Time Savings (VTTS). It is shown for commuting trips that AVs reduce 41% of VTTS on average compared to HVs. For leisure or shopping trips, no significant changes in the VTTS were found. Tscharaktschiew & Evangelinos [23] investigated the impact of the transition in automated driving capabilities (driving mode choice) on road congestion pricing and vice versa, accounting for the interdependencies between traffic flow, the chosen level of autonomous driving, effective road capacity and marginal travel cost. The result suggested that when inconveniences related to autonomous driving are sufficiently high, the imposition of congestion tolls may lead to a situation where drivers abandon autonomous technologies entirely and opt instead for fully manual driving, not the generally expected positive effects.

2.2. AVs Cognition Survey

Along with AVs impact studies, several AVs-related cognition surveys have been conducted. Most cognition studies used surveys, confirming that preference towards AVs was higher for men than for women [24] and higher for the younger generation than for the older [25]. According to Nordhoff et al. [26], most people think that impacts of AVs appear to be positive, but responded that AVs safety benefit needs to be experimentally verified [27]. Moreover, Im et al. [28] analyzed web articles and comments about AVs using text mining techniques, showing that the number of reports that include AVs as a keyword increased, and there are more negative views than positive views. Specifically, the articles and comments with negative views discussed AVs ethics, traffic accidents, and the problem of sudden unintended acceleration.

2.3. AVs Traffic Accidents and Derived AVs Ethics

We reviewed AVs traffic accidents researches and AVs ethics issues for building the design of preventive AVs. Hong et al. [29] & Yang et al. [30] classified types of AVs accident as follows: the negligence of the drivers, accidents due to mechanical defects, malfunction of S/W, and accidents caused by information error, hacking, weather, etc. In the case of the steadily rising trolley dilemma problem [31], doubts were brought up as to whether the trolley dilemma could apply to AVs. They suggested that deriving accident algorithms to respond to it could be misleading [32]. Bae & Lee [33] approached decision-making criteria for the protection priority in accident situations. They proposed two solutions: to enforce them by law and to leave them to the AVs Artificial Intelligence (AI) system itself. In regards to setting them on AVs AI system, Gogoll & Müller [7] discussed differences between Mandatory Ethics Setting (MES), considering society as a whole and Personal Ethics Setting (PES) considering individual interests. For the accident in an ideal condition, Goodall [34] proposed ethical collision algorithms, and Fleetwood [35] discussed Germany's AVs ethics guidelines and main ethical topics.

2.4. Deep Learning Application in Transportation Area

There exist many transportation prediction studies using deep learning technologies. These can be bisected into two categories: prediction of either traffic flow or traffic accidents. First, studies of forecasting traffic flow compare proposed model performance to traditional classical algorithms [36–39]. Particularly, the Long-Short Term Memory (LSTM) model is compared with Statistically Adjusted Engineering (SAE), Radial Basis Function (RBF), Support Vector Machine (SVM), and Auto Regressive Integrated Moving Average (ARIMA) model, and DNNs is compared with Random Forest, a kind of machine learning model. Traffic accident prediction studies were carried out using Social Network Service data (Twitter) using Deep Belief Network (DBN) & LSTM [40]. For real-time accident detection, Chen et al. [41] analyzed the accident impact using GPS-based vehicle data. They used Stack Denoise Autoencoder (SDA), which is more effective to detect accident risk than the traditional models.

Furthermore, as rapid progress is made nowadays in AVs technology, backed by advances in the areas of deep learning and AI, various studies about AVs using AI have been implemented. Especially as AVs requires an accurate perception of surrounding environments to operate reliably, most studies are related to Convolutional Neural Networks (CNNs). That is why object detection is a fundamental function of AVs systems, including camera sensor (2D), Lidar (3D), radar, GPS, etc. Among them, 3D object detection for AVs studies have been carried out recently, and a new methodology (combining or extension) has been proposed. For instance, Li, P., et al. [42] proposed the Stereo R-CNN for 3D object detection as Faster Regions with CNN (Faster R-CNN) for stereo inputs extending to detect objects simultaneously with images on the left and right. The experiments on the challenging KITTI dataset show that their method outperforms the state-of-the-art stereo-based method (Stereo R-CNN) around 30% AP on both 3D detection and 3D localization tasks. Also, Chen, S., et al. [43] developed a CNN–LSTM based on prior knowledge and temporal information for AVs driving. The proposed algorithm was found to be approximately 85% accurate in mimicking human drivers. Zeng, W., et al. [44] proposed the Deep Structured Self-Driving Networks (DSDNet), which performs object detection, motion prediction, and motion planning with a single neural network. The algorithm showed that it has outperformed the state-of-the-art method (DSDNet) on several challenging datasets in general. Existing AVs datasets are limited in the scale and variation of the environments they capture, even though generalization within and between operating regions is crucial to the overall viability of the technology. In response, Sun, P., et al. [45] conducted the study based on the vast amount of real data it owned as a leading group in the study of AVs. This study is based on actual AVs data, and presents a large-scale multimodal camera-LiDAR dataset that is significantly larger, higher quality, and more geographically diverse than any other existing similar dataset. In addition, Djuric, N., et al. [46] introduced a deep learning-based approach that takes into account a current world state and produces raster images of each actor's vicinity for presenting an effective solution to a critical part of the AVs problem. The method first rasterizes actor contexts, followed by training CNNs to use the resulting raster images to predict the actor's short-term trajectory and the corresponding uncertainty. Also, they tested the framework (system), which was deployed to a fleet of AVs.

2.5. Summary

In summary, diverse studies had been conducted about the traffic impacts of AVs. Most studies verified that when AVs are introduced, LOS, speed, and road volume (capacity) are improved and traffic accidents decreased even in a condition of mixed traffic with HVs. Also, cognition studies on AVs social acceptability were carried out. In addition, AI is actively being applied and developed in various ways for AVs. There have been various studies such as AVs control study based on CNNs, End-to-End, traffic volume forecasting based on DNN-Based Traffic Flow prediction (DNN-BTF), and short traffic flow prediction using LSTM.

Although studies about AVs introduction and acceptance surveys have been actively delved into, it is found that there exist limitations in building the foundation for commercialization. In addition, researches on legal and ethical issues on AVs traffic accidents have been carries out from various perspectives, but opinions have been continuously raised whether it is appropriate to answer all the ethical issues related with AVs traffic accidents as pre-conditions of the commercialization. Similarly, most AVs technology researches have gradually advanced to study related to the technology supplementation of themselves, but the researches related with the prevention of AVs traffic accidents have been found insufficient. Therefore, as proposed in the German and U.S. ethics guidelines and NVIDIA report [47], this study aims to lay the foundation for the design of the PADS using AI as a way to prevent AVs traffic accidents in advance.

3. Methodology

DNNs we intend to use in this study are a kind of Artificial Neural Network, consisting of the input layer, the output layer, and the hidden layers in between. DNNs are capable of modeling complex non-linear relationships, such as common artificial neural networks, with the ability to express basic elements in hierarchical configurations and the added layers to converge the characteristics of lower layers. In addition, regardless of continuous or categorical variables, non-linear combinations between input variables are easy to analyze, and automatic feature extraction reduces the hassle of variable selection. These features are used in the study to extract ambient situation information factors, such as weather information, external factors, etc. [39].

Traffic accidents are issues directly related to human life, and it is believed that legal and ethical problems will be inevitable in the event of an accident by learning and predicting inaccurate content. Thus, the German Ethics Guidelines for AVs stated that they should be designed to prevent accidents in advance and that they allow the use of AI technology to improve safety. Therefore, it is considered that the top priority is to learn how to prevent accidents by recognizing accident situations in advance as a solution to traffic accidents of AVs, and in this study, DNN, which has higher predictability than conventional machine learning algorithms, is to be used.

3.1. Data Collecting & Pre-Process

The data used for this study was from the accident data of Seoul city collected from 2017 to 2018. The main dataset is Traffic Accident Analysis System (TAAS) [48] data provided by The Road Traffic Authority (KoROAD). The information on traffic accident conditions in TAAS can be obtained in the form of Excel data on the TAAS website. However, it was necessary to extract the location information (coordinates) of an accident to identify the traffic situation information such as several lanes, speed, etc. during the traffic accident in detail. So, we crawled location coordinates data from TAAS to merge link attributes with traffic accident condition data. The crawled data include location coordinates, accident number, date of the accident, day of the accident, the content of accident, the number of deaths, the number of severe injuries, the number of light injuries, the number of wounded, accident type, violation of the law, road conditions, weather, road type, offender information, and victim information (see Table 1).

Table 1. Crawled Traffic Accident Analysis System (TAAS) data.

Class	Feature	Class	Feature	Class	Feature
A_NO	Accident Assignment Number	WO_NO	The number of Wounded	O_sex	Offender Sex
Date	Date of Accident(YY/MM/DD)	A_type	Accident Type	O_age	Offender Age
Day of the week	Day of Accident	V_Law	Violation of Law	O_injury	Offender injury degree
A_fact	Content of Accident	R_con	Road Conditions	V_car	Victim Car
D_NO	The number of Death	W_con	Weather	V_sex	Victim Sex
SI_NO	The number of Severe injuries	R_type	Road Type	V_age	Victim Age
LI_NO	The number of light injuries	O_car	Offender Car	V_injury	Victim injury degree

However, TAAS has a limitation in that it does not provide data on traffic environment such as the number of lanes and speed, etc. Accordingly, we used Transport Operation & Information Service (TOPIS) [49] from the Seoul city and Korea Transport DataBase (KTDB) [50] provided by MOLIT to get the traffic environment data of each node & link. The link speed data were drawn from TOPIS (See Table 2), and the data on the number of lanes were extracted from KTDB (node-link data). TOPIS and KTDB data were obtained and utilized in Excel form from the above-mentioned sites.

Table 2. Sample of TOPIS speed data.

Date	Day of the Week	Link ID	Road Type	Time		
				1	2	3
20180601	Fri.	1080012200	Minor arterial road	44.6	27.88	51.79
20180601	Fri.	1080012800	Minor arterial road	17.73	24.32	24.67
20180601	Fri.	1080012700	Minor arterial road	23.15	29.9	32.56
20180601	Fri.	1080012100	Minor arterial road	47.88	40.4	44.29
20180601	Fri.	1230024700	Other road	21.62	21.17	27.13
20180601	Fri.	1230019500	Other road	30.59	28.86	30.7

TOPIS provides link speed data only on minor arterial roads, but not on collector roads. Thus, we utilized KTDB to get the data for collector roads. The data on speed and number of lanes were merged based on the TAAS coordinate system, and we used the 1-h data for the learning.

After the collection, we refined the data for the analysis. We excluded X, Y coordinates, local area name, and accident number from the dataset because they are only useful for merging purposes. Also, the data on the number of casualties such as the number of deaths, severe injuries, etc. were excluded because they are deemed unsuitable for this study which aims to prevent accidents in advance. When checking the basic statistics of the data, it was found that the number of accidents was appeared constant in monthly and daily bases. Also, the seasonality that we wanted to check in monthly accidents seems to be well reflected in the "weather" factor; likewise for the "day" factor reflected by the "day of the week" factor. Therefore, we only used the time and the day of the week data in the analysis.

Since TOPIS data do not include speed data under the minor arterial road level, 10 to 20 km/h, which is the average speed in Seoul, was allocated for the empty data cells. Finally, a total of 77,000 pre-processed data were used, with 38,625 cases in 2017 and 38,796 cases in 2018.

3.2. Learning Process

The purpose of this study is to prevent AVs-related accidents in advance by learning from the accident analysis of HVs. The dependent variables adopted for offender and victim identically consist of 5 different degrees of injuries such as death, serious injury, injury, minor injury, and no injury. Traffic accidents do not always consist of two parties, the offender and the victim. Sometimes there exists only one party case. For instance, when only vehicle damage is occurred with no human casualties (only one party), there exist only offenders. These cases also are incorporated into the learning process. The independent variables accounting for the seriousness of the injuries consist of road conditions, weather, road shape, number of lanes, and link speed (See Table 3). The learning was implemented by adding new factors such as time (T), day of the week (D), vehicle type of offender & victim (C), and violation of the law (L) in a step-by-step manner.

Table 3. Basic Data Variable Setting for Learning.

Feature	Class	Categorical Variable
Offender injury degree	Output Layer	1~6
Victim injury degree	Output Layer	1~8
Road Condition (RC)	Input Layer	1~5
Weather (W)	Input Layer	1~5
Road shape (S)	Input Layer	1~4
Number of Lane (L)	Input Layer	0~5
Link Speed (Sp)	Input Layer	1~6

4. Design of Optimal Deep Neural Networks (DNNs)

This chapter proposes an optimal DNNs for more accurate traffic accident prediction. The process of optimizing the model consists of three steps: setting up the data range—epoch—and hidden layers.

4.1. Environment on Building Algorithm

This study intends to produce an optimal model of accident prediction based on the backpropagation algorithm [51] and SGD (Stochastic Gradient Descent) optimizer of the Tensorflow and Keras libraries [52,53]. The ratio of training and testing data was set at 8:2. It used the dropout of randomly skipping a certain amount between nodes in weight update for minimizing an overfitting. We experimentally confirmed that the highest prediction was achieved where the dropout value was set at 0.2. Also, we used ReLU (Rectified Linear Unit) function [54] to prevent gradient vanishing when using the Sigmoid function [55]. The batch size was set to 64 for stable learning.

4.2. Learning Data Range for Building optimal DNNs

First, we performed DNNs using 2017, 2018, and 2017–2018 integrated data. For extracting optimal data range suited to the model, we set up the default model (RC, W, S, L, Sp; ① set, which are only external factors) and simulated the degree of injury to the offender and victim as output. The prediction results show that the predicted accuracy of the degree of injury to the offender is over 80%. Also, it appeared that the 2018 data's prediction accuracy was higher by about 0.1–2.0% than the 2017 data, and the 2017–2018 integrated data was more accurate by about 0.5–1.5% than the 2018 data. The degree of injury to the victim showed about 60–65% accuracy, and prediction based on the integrated data was found more accurate by 0.5–2.5% than 2017, 2018 data. Therefore, we decide to implement the simulation by using the 2017–2018 integrated data set (See Table 4 & Figure 1).

Table 4. Offender/Victim Injuries Accuracy by datasets.

Feature	DATA					
	Offender			Victim		
	2017	2018	2017–2018	2017	2018	2017–2018
①Set	80.80	81.20	81.16	57.99	59.89	61.05
①Set + T	80.60	81.80	81.32	57.22	59.41	61.21
①Set + D	80.16	80.82	80.85	58.79	59.51	60.75
①Set + T + D	79.98	82.00	80.93	58.55	59.77	60.92
①Set + C	82.26	83.96	84.02	64.12	65.36	66.46
①Set + L	81.22	80.77	81.41	58.86	59.41	60.67
①Set + T + D + C	81.52	82.88	81.89	64.16	65.12	66.43
①Set + T + D + L	80.89	81.08	84.12	58.11	60.26	62.75
①Set + C + L	83.04	83.15	84.38	64.07	65.01	65.83
All data	83.37	82.74	84.15	62.92	65.56	65.98

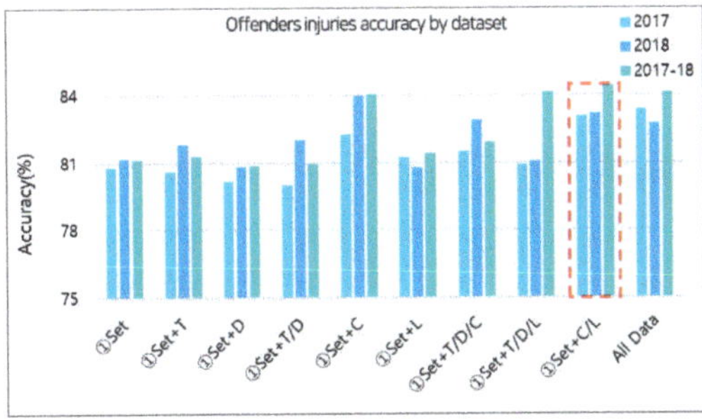

(**a**) Offender injuries accuracy

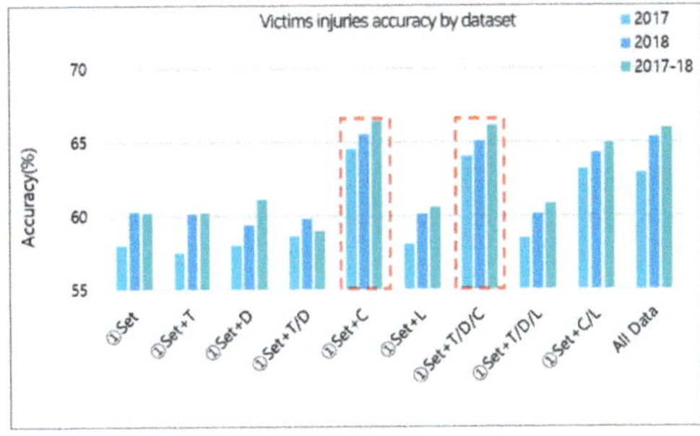

(**b**) Victim injuries accuracy

Figure 1. (**a**) Offender Injuries Accuracy by dataset; and (**b**) Victim Injuries Accuracy by dataset: The red dash line rectangular stands for the highest accuracy at that factors in 2017–2018

4.3. Learning Data Epoch for Building Optimal DNNs

Second, we performed DNNs by varying epoch, a state where one learning is completed for the entire data set within the network. When the epoch is set up on a large scale, training can cause overfitting, resulting in less accuracy in testing, verification, and application of new data. So, the step equally simulated one step (data range) process for extracting optimal data epoch. We set up epoch in five divisions of 100 units and simulated the degree of injury to the offender and victim as output. It is confirmed that the highest accuracy was achieved in both offender and victim at 100 epochs. The result suggests that the higher epoch was set up, the lower accuracy was shown due to overfitting. Therefore, we will simulate the study by setting 2017–2018 integrated data and 100 epochs (See Table 5).

Table 5. Offender/Victim Injuries Accuracy by Epoch.

Feature	Epoch									
	Offender					Victim				
	100	200	300	400	500	100	200	300	400	500
①Set	81.16	81.02	80.89	81.13	81.17	61.05	60.55	60.87	90.13	60.75
①Set + T	81.32	81.20	81.22	80.98	81.25	61.21	61.13	60.75	60.81	60.92
①Set + D	80.85	80.80	80.90	80.01	80.57	60.75	60.43	61.21	60.84	61.33
①Set + T + D	80.93	80.77	80.89	80.65	80.91	60.92	60.88	60.13	60.74	60.27
①Set + C	84.02	83.70	83.52	83.78	83.97	66.60	65.43	64.79	66.60	65.98
①Set + L	81.41	81.50	81.13	80.87	80.15	60.67	59.70	60.12	60.45	60.33
①Set + T + D + C	81.89	81.73	81.51	81.27	81.55	66.43	65.47	66.11	65.87	65.90
①Set + T + D + L	84.12	83.96	83.87	83.17	84.03	62.75	61.72	62.70	62.13	62.60
①Set + C + L	84.38	84.27	84.29	84.02	84.30	65.83	64.98	65.74	65.10	65.33
All data	84.15	84.05	83.79	83.98	84.03	65.98	64.80	64.93	65.77	65.68

4.4. Setting the Hidden Layers Building Optimal DNNs

Third, we performed DNNs by setting the hidden layers and the number of nodes. The complexity of neural networks is determined by these two settings. It is important to set up the optimal node and hidden layers suitable to the model because model overfitting might occur and lead to poor learning. So, this means that the hidden layers and the number of nodes should be set up to suit this model.

For obtaining the optimal level of hidden layers and nodes, we tried to learn based on ① Set. However, the results showed that the learning was not done properly due to a shortage in the number of features in ① Set. Thus, we extracted a hidden layer based on "all data" which contained a default feature and new factors (Time, Day, Car of offender & victim, violation of Law).

Consequently, the prediction accuracy of injury to the offender was observed to be mostly higher than 84%, and the highest accuracy was 84.15% with (256,128,64,64) nodes in the hidden layers. The prediction accuracy of injury to the victim reached mostly higher than 64%, and the highest accuracy was 65.93% with (256,128,64,64) nodes in hidden layers as well. It suggests that the optimal hidden layers about traffic accident prediction is (256,128,64,64), which deduced output value to the converging process. Therefore, we will simulate the study by setting 2017–2018 integrated data, 100 epochs, and (256,128,64,64) hidden layers(See Figure 2 & Table 6).

Table 6. Offender/Victim Injuries Accuracy by Hidden Layers.

The Number of Nodes in Hidden Layer	Accuracy	
	Offender	Victim
(256,256,128,64)	84.10	64.21
(256,128,128,64)	83.97	65.02
(256,128,64,64)	4.15	65.93
(256,128,64,32)	84.06	64.38
(128,128,64,64)	83.87	64.50
(128,128,64,32)	84.08	64.32
(128,64,64,32)	84.05	64.25
(128,64,64)	83.91	64.24
(128,64,32)	83.82	64.21
(64,64,64)	84.12	64.31

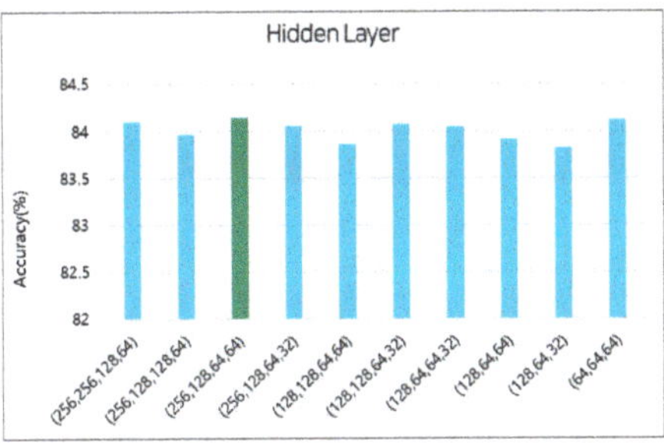

(**a**) Offender injuries accuracy

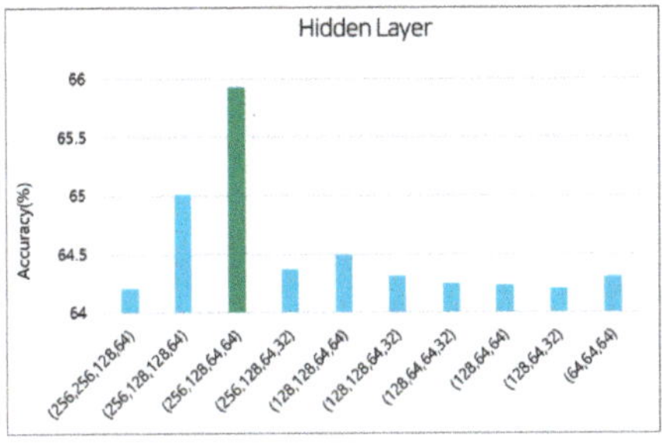

(**b**) Victim injuries accuracy

Figure 2. (**a**) Offender Injuries Accuracy by Hidden Layers; and (**b**) Victim Injuries Accuracy by Hidden Layers: Different colored bars stand for the highest accuracy is derived from the hidden layers.

5. Result

5.1. Extracting Traffic Accidents Context Fatures

Learning based on the optimization model was carried out to identify factors affecting the degree of offender injury. As shown in Table 7 and Figure 4, the ① set, which only uses external factors, showed 81.16% prediction accuracy. It means that more than 80% of the injury to the offender can be predicted using external factors only. In addition, 84% prediction accuracy was achieved on the degree of injuries if additional factors of the vehicle were applied to the ① set, and 84.38% accuracy was confirmed when the vehicle type and the violation of the law were taken into consideration. As with offender factors, 61.05% accuracy was confirmed for the victim's injury when using external factors only. In addition, 66.46% of accuracy was observed in predicting victim's injuries when vehicle type was added to ① Set and 64.43% accuracy when adding time and day factors along with the vehicle type.

As a result of the analysis of the degree of injury of the offender, high accuracy is confirmed when the vehicle type and the violation of the law are added to the speed, the external factor. In addition, when the time, day, and vehicle types are set as additional factors in addition to external factors, high accuracy is achieved in the analysis of the victim's injury. The time series and vehicle type are considered to be the main factors in determining the victim's injury (See Table 7 & Figure 3).

Following the learning through the optimization model, we identified specific features that determine the degree of traffic accident injury. Based on the analysis through DNN learning, a random forest feature importance analysis is performed, which is a machine learning technique, to confirm factors more accurately. As a result, the main factors of injuring offenders were identified in the order of "Vehicle Type, Speed, Time, and Day of the week", and those of the victims were in the order of "Speed, Time, Victim Vehicle Type, and Day of the Week". It was found that the factors that determine the degree of injury to the offender and the victim are similar. However, for the offender, the vehicle type is found more critical, while for the victim, the link speed is identified as more important (See Table 8 & Figure 4).

Table 7. Offender/Victim Injuries Prediction Accuracy using optimal DNNs model.

Feature	Accuracy	
	Offender	Victim
①Set	81.16	61.05
①Set + T	81.32	61.21
①Set + D	80.85	60.75
①Set + T + D	80.93	60.92
①Set + C	84.02	66.46
①Set + L	81.41	60.67
①Set + T + D + C	81.89	66.43
①Set + T + D + L	84.12	62.75
①Set + C + L	84.38	65.83
All data	84.15	65.98

(**a**) Offender injuries accuracy

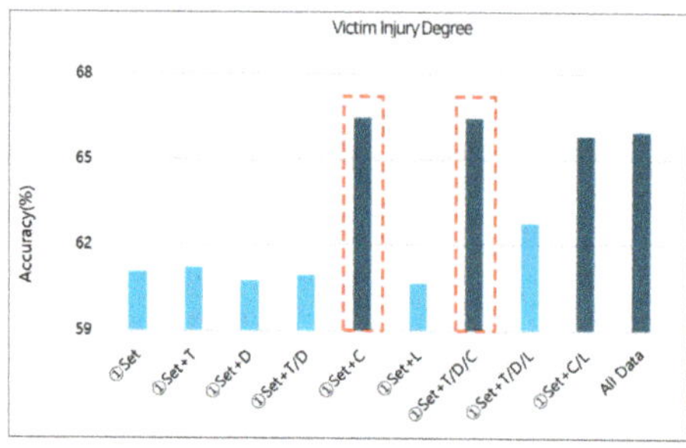

(**b**) Victim injuries accuracy

Figure 3. (**a**) Offender Injuries Accuracy by Optimal DNNs; (**b**) Victim Injuries Accuracy by Optimal DNNs: The red dash line rectangular stands for the highest accuracy at that factors.

Table 8. Result of Importance Analysis for Traffic Accident Factors.

Feature	Accuracy									
Offender Factors	Offender Vehicle Type (29.8)	Speed (18.9)	Time (15.5)	Day of the Week (9.4)	Violation of the Law (7.2)	Number of Lane (6.6)	Victim Vehicle Type (6.1)	Road Shape (3.6)	Weather (1.6)	Road Condition (1.4)
Victim Factors	Speed (22.5)	Time (18.4)	Victim Vehicle Type (12.9)	Day of the Week (10.9)	Offender Vehicle Type (29.8)	Number of Lane (6.6)	Violation of the Law (7.2)	Road Shape (3.6)	Weather (1.6)	Road Condition (1.4)

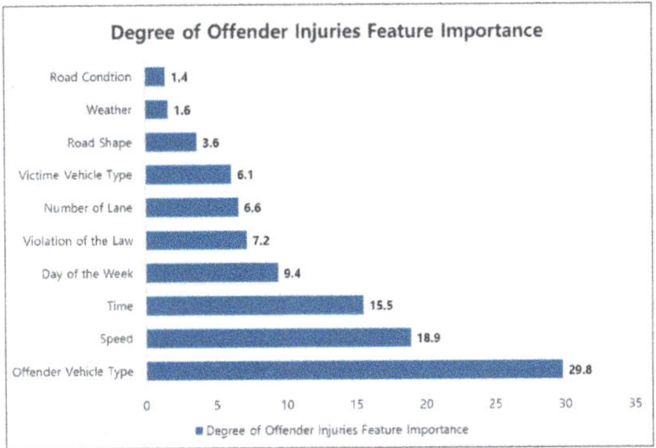

(**a**) Offender injuries Feature Importance

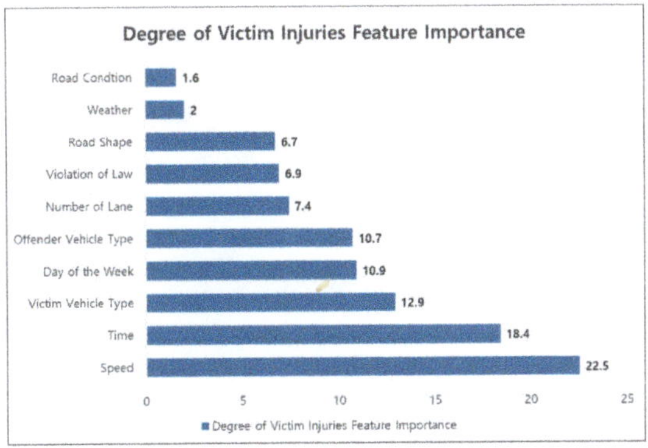

(**b**) Victim injuries Feature Importance

Figure 4. (**a**) Injuries Feature Importance; (**b**) Victim Injuries Feature Importance

5.2. Additional Consideration

Although the main factors determining the degree of offender and victim injury were identified with this study, it is necessary to find out why there exist the differences in accuracy between the two, notwithstanding the same dataset were used for the analysis. It is because the principal cause of injury to the offender is the vehicle type whose data was originated from the TAAS that requires no data correction, but the cause of injury to the victim (speed) is the data obtained from the TOPIS data that entail a lot of missing values (see Section 3.1). It is judged that the difference in the factors that determine the injury occurred in the original data (TAAS & TOPIS), lead to the differences in accuracy. However, the accuracy derived from this analysis is higher than that of existing statistics-based traffic accident prediction models (Poisson regression model, negative binomial regression model, etc. [56–59]. This is judged to be meaningful, as car accidents can be prevented in advance by predicting traffic accident injuries through DNN learning.

6. Conclusions

Automated Vehicles are under the spotlight as an alternative to diminishing traffic accidents. However, AVs accidents that occurred during several test drives led to worry about AVs safety. When AVs are commercialized and mixed with human driving vehicles, AVs-related accidents can occur even under ideal conditions. In order to solve this problem, studies on AVs ethics have been conducted, but it was difficult to obtain an adequate answer. So, Germany and the United States issued AVs ethical guidelines emphasizing the prevention of traffic accidents. This study analyzed the traffic accident data based on DNNs to propose preventive measures for AVs accidents.

For the analysis of traffic accidents, we preprocessed TAAS, standard node-link data, and TOPIS data. In the DNNs analysis, the input layer consists of external factors and additional factors, and the output layer was set to the degree of injuries for both offenders & victims. Also, to construct the optimal hidden layers, we controlled the learning data range, the epoch, and the number of nodes in the hidden layers. For the analysis, we conducted learning by adding input factors incrementally. The results show that in the offender case, 81% of prediction accuracy was achieved when only external factors were considered, while the accuracy increased to 85% when factors such as violation of law and vehicle types are added to the analysis. In the victim case, the prediction accuracy remained at 61% with external factors only, but the accuracy increased to 67% when additional factors were taken into consideration. The main reason determining about 20% prediction accuracy deviation between offender and victim comes from the level of data accuracy of the key injury determinant between offender and victim. The offender's degree of injury is largely determined by the type of vehicle, which did not require the data collection. On the other hand, the speed which affected the degree of victim injury most were from the merged data of TOPIS and TAAS with many missing data. The analysis revealed that factors such as vehicle type, time, and day were found important in determining the degree of injury. In addition, a random forest importance analysis identified that the injury determinants of the offender were ordered vehicle type, speed, time, and day, while those of victims were speed, time, vehicle type, and day.

As discussed, vehicle type and speed were identified as the main factors of determining injury, respectively. Accordingly, in the future, we plan to conduct researches that can more accurately prevent accidents by recognizing vehicle types using the CNNs (YOLO, etc.) technique, and that can predict the degree of injury according to speed or suggesting an appropriate vehicle safety distance. Also, future study will apply a developed methodology (Meta AI etc.) using updated traffic accident data. However, since data from HVs were utilized to develop preventive measures for AVs accidents, there is a limit to directly apply the results to AVs. Nevertheless, since the analysis focused on external factors (such as weather, road conditions, road types, etc.) that are difficult to control even AVs, it is expected that they will be used as basic data for analyzing the impact of external factors related to traffic accidents related to AVs in the future. Future study is needed to analyze traffic accidents by using data on the causes (e.g., weather, body defects, etc.) of AVs accidents as input values.

Author Contributions: Conceptualization, K.H.; Data curation, J.S.; Formal analysis, M.K.; Investigation, M.K.; Methodology, M.K.; Project administration, K.H.; Software, M.K. and J.S.; Visualization, M.K.; Writing—original draft, M.K.; Writing—review & editing, M.K. and K.H. All authors have read and agreed to the published version of the manuscript.

Funding: This paper was supported by a grant (code 20TLRP-B148970-03) from Transportation and Logistics R&D Program (TLRP) funded by Ministry of Land, Infrastructure and Transport of Korean government.

Acknowledgments: This paper is a revised and complemented content of "A Study on DNN (Deep Neural Network)-based Traffic Accident Context Analysis for The Design of Preventive Automated Driving System" Minhee Kang's Master Degree.

Conflicts of Interest: The authors declare no conflict of interest.

References

1. SAE On-Road Automated Vehicle Standards Committee. *Taxonomy and Definitions for Terms Related to Driving Automation Systems for on-Road Motor Vehicles*; SAE International: Warrendale, PA, USA, 2018.
2. National Highway Traffic Safety Administration. *Vision for Safety 2.0 Guidance for Automated Vehicles*; US Department of Transportation: Washington, DC, USA, 2017.
3. Lee, K.; Jeon, S.; Kim, H.; Kum, D. Optimal path tracking control of autonomous vehicle: Adaptive full-state linear quadratic gaussian (lqg) control. *IEEE Access* **2019**, *7*, 109120–109133. [CrossRef]

4.   Singh, S. *Critical Reasons for Crashes Investigated in the National Motor Vehicle Crash Causation Survey (No. DOT HS 812 115)*; US Department of Transportation: Washington, DC, USA, 2015.

5.   Abraham, H.; Lee, C.; Brady, S.; Fitzgerald, C.; Mehler, B.; Reimer, B.; Coughlin, J.F. Autonomous vehicles and alternatives to driving: Trust, preferences, and effects of age. In Proceedings of the transportation research board 96th annual meeting (TRB'17), Washington, DC, USA, 8–12 January 2017.

6.   Zhang, T.; Tao, D.; Qu, X.; Zhang, X.; Lin, R.; Zhang, W. The roles of initial trust and perceived risk in public's acceptance of automated vehicles. *Transp. Res. Part C Emerg. Technol.* **2019**, *98*, 207–220. [CrossRef]

7.   Hartwich, F.; Witzlack, C.; Beggiato, M.; Krems, J.F. The first impression counts—A combined driving simulator and test track study on the development of trust and acceptance of highly automated driving. *Transp. Res. Part F Traffic Psychol. Behav.* **2019**, *65*, 522–535. [CrossRef]

8.   Gogoll, J.; Müller, J.F. Autonomous Cars: In Favor of a Mandatory Ethics Setting. *Sci. Eng. Ethics* **2016**, *23*, 681–700. [CrossRef] [PubMed]

9.   Gurney, J.K. Crashing into the unknown: An examination of crash-optimization algorithms through the two lanes of ethics and law. *Albany Law Rev.* **2015**, *79*, 183.

10. Bonnefon, J.F.; Shariff, A.; Rahwan, I. Autonomous vehicles need experimental ethics: Are we ready for utilitarian cars? *arXiv* **2015**, arXiv:1510.03346.

11. Federal ministry of transport and digital infrastructure. *Ethics Commission: Automated and Connected Driving*; BMVI: Berlin, Germany, 2017.

12. National Highway Traffic Safety Administration. *Federal Automated Vehicles Policy: Accelerating the Next Revolution in Roadway Safety*; US Department of Transportation: Washington, DC, USA, 2016.

13. Choi, J.K.; Ji, Y.G. Investigating the Importance of Trust on Adopting an Autonomous Vehicle. *Int. J. Hum. Comput. Interact.* **2015**, *31*, 692–702. [CrossRef]

14. Xu, Z.; Zhang, K.; Min, H.; Wang, Z.; Zhao, X.; Liu, P. What drives people to accept automated vehicles? Findings from a field experiment. *Transp. Res. Part C Emerg. Technol.* **2018**, *95*, 320–334. [CrossRef]

15. Man, S.S.; Xiong, W.; Chang, F.; Chan, A.H.S. Critical Factors Influencing Acceptance of Automated Vehicles by Hong Kong Drivers. *IEEE Access* **2020**, *8*, 109845–109856. [CrossRef]

16. Ko, Y.; Rho, J.H.; Donghyung, Y. Assessing Benefits of Autonomous Vehicle System Implementation through the Network Capacity Analysis. *Korea Spat. Plan. Rev.* **2017**, *93*, 17–24. [CrossRef]

17. Tibljaš, A.D.; Giuffrè, T.; Surdonja, S.; Trubia, S. Introduction of Autonomous Vehicles: Roundabouts Design and Safety Performance Evaluation. *Sustainability* **2018**, *10*, 1060. [CrossRef]

18. Wang, Y.; Wang, L. Autonomous vehicles' performance on single lane road: A simulation under VISSIM environment. In Proceedings of the 2017 10th International Congress on Image and Signal Processing, BioMedical Engineering and Informatics (CISP-BMEI) IEEE, Shanghai, China, 14–16 October 2017; pp. 1–5.

19. Kang, M.H.; Im, I.J.; Song, J.I.; Hwang, K.Y. Analyzing Traffic Impacts of Automated Vehicle on Expressway Weaving Sections—A Case Study using Seoul-Singal Ramp Area-. *J. Transp. Res.* **2019**, *26*, 59–75. (In Korean)

20. Lee, S.Y.; Oh, M.S.; Oh, C.; Jeong, E.B. Automated Driving Aggressiveness for Traffic Management in Automated Driving Environments. *J. Korean Soc. Transp.* **2018**, *36*, 38–50. (In Korean) [CrossRef]

21. Morando, M.M.; Tian, Q.; Truong, L.T.; Vu, H.L. Studying the Safety Impact of Autonomous Vehicles Using Simulation-Based Surrogate Safety Measures. *J. Adv. Transp.* **2018**, *2018*. [CrossRef]

22. Kolarova, V.; Steck, F.; Bahamonde-Birke, F.J. Assessing the effect of autonomous driving on value of travel time savings: A comparison between current and future preferences. *Transp. Res. Part A Policy Pract.* **2019**, *129*, 155–169. [CrossRef]

23. Tscharaktschiew, S.; Evangelinos, C. Pigouvian road congestion pricing under autonomous driving mode choice. *Transp. Res. Part C Emerg. Technol.* **2019**, *101*, 79–95. [CrossRef]

24. König, M.; Neumayr, L. Users' resistance towards radical innovations: The case of the self-driving car. *Transp. Res. Part F Traffic Psychol. Behav.* **2017**, *44*, 42–52. [CrossRef]

25. Abraham, H.; Lee, C.; Brady, S.; Fitzgerald, C.; Mehler, B.; Reimer, B.; Coughlin, J.F. *Autonomous Vehicles, Trust, and Driving Alternatives: A Survey of Consumer Preferences*; MIT AgeLab: Cambridge, MA, USA, 2016; pp. 1–16.

26. Nordhoff, S.; Van Arem, B.; Happee, R. Conceptual Model to Explain, Predict, and Improve User Acceptance of Driverless Podlike Vehicles. *Transp. Res. Rec. J. Transp. Res. Board* **2016**, *2602*, 60–67. [CrossRef]

27. Kaan, J. User Acceptance of Autonomous Vehicles. Master's Thesis, Faculty of Technology, Policy and Management (TPM), Delft University of Technology, Delft, The Netherlands, 2017.

28. Im, I.J.; Song, J.I.; Lee, J.Y.; Hwang, K.Y. Analysis of the Perception of Autonomous Vehicles Using Text Mining Technique. *J. Korea Inst. Intell. Transp. Syst.* **2017**, *16*, 231–243.

29. Hong, T.-S.; Kwon, Y.-S.; Kuack, B.-S.; Lee, J.-C. An Exploratory Study on Criminal Responsibility of Self-driving Car Accident Type. *Leg. Theory Pract. Rev.* **2018**, *6*, 231–252. [CrossRef]

30. Yang, J.S.; Kim, K.J.; Kwon, H.B. A Study on the Differences of Perception on the Causes of Self-driving Cars in Human Distribution. *J. Distrib. Manag. Res.* **2019**, *22*, 15–22. (In Korean)

31. Thomson, J.J. The Trolley Problem. *Yale Law J.* **1985**, *94*, 1395. [CrossRef]

32. Kim, S.-R. The Ethical Approach to Traffic Accident-Algorithm of Autonomous Driving Vehicle and Its Legal Implications. *IT LAW Rev.* **2017**, *15*, 193–219. [CrossRef]

33. Bae, I.H.; Lee, K.H. Criteria of the Prior Protection in Fully Autonomous Car Accidents. *Study Future* **2018**, *3*, 25–48. (In Korean)

34. Goodall, N.J. Ethical Decision Making during Automated Vehicle Crashes. *Transp. Res. Rec. J. Transp. Res. Board* **2014**, *2424*, 58–65. [CrossRef]

35. Fleetwood, J. Public Health, Ethics, and Autonomous Vehicles. *Am. J. Public Health* **2017**, *107*, 532–537. [CrossRef] [PubMed]

36. Wu, Y.; Tan, H.; Qin, L.; Ran, B.; Jiang, Z. A hybrid deep learning based traffic flow prediction method and its understanding. *Transp. Res. Part C Emerg. Technol.* **2018**, *90*, 166–180. [CrossRef]

37. Polson, N.G.; Sokolov, V.O. Deep learning for short-term traffic flow prediction. *Transp. Res. Part C Emerg. Technol.* **2017**, *79*, 1–17. [CrossRef]

38. Zhao, Z.; Chen, W.; Wu, X.; Chen, P.C.Y.; Liu, J. LSTM network: A deep learning approach for short-term traffic forecast. *IET Intell. Transp. Syst.* **2017**, *11*, 68–75. [CrossRef]

39. Kim, D.H.; Hwang, K.Y.; Yoon, Y. Prediction of Traffic Congestion in Seoul by Deep Neural Network. *J. Korea Inst. Intell. Transp. Syst.* **2019**, *18*, 44–57. [CrossRef]

40. Zhang, Z.; He, Q.; Gao, J.; Ni, M. A deep learning approach for detecting traffic accidents from social media data. *Transp. Res. Part C Emerg. Technol.* **2018**, *86*, 580–596. [CrossRef]

41. Zhang, W.; Liu, K.; Zhang, W.; Zhang, Y.; Gu, J. Deep Neural Networks for wireless localization in indoor and outdoor environments. *Neurocomputing* **2016**, *194*, 279–287. [CrossRef]

42. Li, P.; Chen, X.; Shen, S. Stereo R-CNN Based 3D Object Detection for Autonomous Driving. In Proceedings of the IEEE Conference on Computer Vision and Pattern Recognition, Long Beach, CA, USA, 16–20 June 2019; pp. 7636–7644.

43. Chen, S.; Leng, Y.; Labi, S. A deep learning algorithm for simulating autonomous driving considering prior knowledge and temporal information. *Comput. Civ. Infrastruct. Eng.* **2019**, *35*, 305–321. [CrossRef]

44. Zeng, W.; Wang, S.; Liao, R.; Chen, Y.; Yang, B.; Urtasun, R. DSDNet: Deep Structured self-Driving Network. *arXiv* **2020**, arXiv:2008.06041.

45. Sun, P.; Kretzschmar, H.; Dotiwalla, X.; Chouard, A.; Patnaik, V.; Tsui, P.; Guo, J.; Zhou, Y.; Chai, Y.; Caine, B.; et al. Scalability in Perception for Autonomous Driving: Waymo Open Dataset. In Proceedings of the IEEE/CVF Conference on Computer Vision and Pattern Recognition, Seattle, WA, USA, 13–19 June 2020; pp. 2443–2451.

46. Djuric, N.; Radosavljevic, V.; Cui, H.; Nguyen, T.; Chou, F.-C.; Lin, T.-H.; Singh, N.; Schneider, J. Uncertainty-aware Short-term Motion Prediction of Traffic Actors for Autonomous Driving. In Proceedings of the IEEE Winter Conference on Applications of Computer Vision, Snowmass Village, CO, USA, 1–5 March 2020; pp. 2095–2104.

47. NVIDIA. *NVIDIA Self-Driving Safety Report*; NVIDIA: Santa Clara, CA, USA, 2018.

48. Traffic Accident Analysis System. 2019. Available online: http://taas.koroad.or.kr (accessed on 20 September 2020).

49. Seoul Transport Operation and Information Services. 2018. Available online: http://topis.seoul.go.kr (accessed on 20 September 2020).

50. Korea Transport DataBase. 2019. Available online: https://www.ktdb.go.kr (accessed on 20 September 2020).

51. Rumelhart, D.E.; Hinton, G.E.; Williams, R.J. Learning representations by back-propagating errors. *Nature* **1986**, *323*, 533. [CrossRef]

52. TensorFlow. Available online: https://www.tensorflow.org/ (accessed on 20 September 2020).

53. Keras. Available online: https://keras.io/ (accessed on 20 September 2020).

54. Nair, V.; Hinton, G.E. Rectified linear units improve restricted boltzmann machines. In Proceedings of the 27th International Conference on Machine Learning (ICML-10), Haifa, Israel, 21–24 June 2010; pp. 807–814.

55. Bridle, J.S. Training stochastic model recognition algorithms as networks can lead to maximum mutual information estimation of parameters. In *Advances in Neural Information Processing Systems 2*; Morgan Kaufmann Publishers Inc.: San Francisco, CA, USA, 1990; pp. 211–217.

56. Lee, S.H.; Lee, Y.D.; Do, M.S. Safety Impacts of Red Light Enforcement on Signalized Intersections. *J. Korean Soc. Transp.* **2012**, *30*, 93–102. (In Korean) [CrossRef]

57. Lee, S.H.; Park, M.H.; Woo, Y.H. A Study on Development Crash Prediction Model for Urban Intersections Considering Random Effects. *J. Korea Inst. Intell. Transp. Syst.* **2015**, *14*, 85–93. (In Korean) [CrossRef]

58. Lee, S.H.; Woo, Y.H. Safety Performance Functions for Central Business Districts Using a Zero-Inflated Model. *Int. J. Highw. Eng.* **2016**, *18*, 83–92. (In Korean) [CrossRef]

59. Lee, S.H.; Lee, Y.D.; Do, M.S. Analysis on Safety Impact of Red Light Cameras using the Empirical Bayesian Approach. *Transp. Lett. Int. J. Transp. Res.* **2016**, *8*, 241–249. [CrossRef]

 *electronics*

Article

An Application of Reinforced Learning-Based Dynamic Pricing for Improvement of Ridesharing Platform Service in Seoul

Jaein Song [1], Yun Ji Cho [2], Min Hee Kang [2] and Kee Yeon Hwang [3,*

[1]   Research Institute of Science and Technology, Hongik University, Seoul 04066, Korea; wodlsthd@nate.com
[2]   Department of Smart City, Hongik University, Seoul 04066, Korea; yunji9943@gmail.com (Y.J.C.);
      speakingbee@hanmail.net (M.H.K.)
[3]   Department of Urban Planning, Hongik University, Seoul 04066, Korea
*   Correspondence: keith@hongik.ac.kr

Received: 29 September 2020; Accepted: 30 October 2020; Published: 2 November 2020

Abstract: As ridesharing services (including taxi) are often run by private companies, profitability is the top priority in operation. This leads to an increase in the driver's refusal to take passengers to areas with low demand where they will have difficulties finding subsequent passengers, causing problems such as an extended waiting time when hailing a vehicle for passengers bound for these regions. The study used Seoul's taxi data to find appropriate surge rates of ridesharing services between 10:00 p.m. and 4:00 a.m. by region using a reinforcement learning algorithm to resolve this problem during the worst time period. In reinforcement learning, the outcome of centrality analysis was applied as a weight affecting drivers' destination choice probability. Furthermore, the reward function used in the learning was adjusted according to whether the passenger waiting time value was applied or not. The profit was used for reward value. By using a negative reward for the passenger waiting time, the study was able to identify a more appropriate surge level. Across the region, the surge averaged a value of 1.6. To be more specific, those located on the outskirts of the city and in residential areas showed a higher surge, while central areas had a lower surge. Due to this different surge, a driver's refusal to take passengers can be lessened and the passenger waiting time can be shortened. The supply of ridesharing services in low-demand regions can be increased by as much as 7.5%, allowing regional equity problems related to ridesharing services in Seoul to be reduced to a greater extent.

Keywords: dynamic pricing; reinforcement learning; ridesharing; supply improvement; taxi

---

1. Introduction

Since the 1960s, with the beginning of industrialization, Korea experienced a rapid urbanization [1–3]. As such, the transportation infrastructure has also developed steadily, greatly improving the mobility of citizens. However, in metropolitan areas, such as the capital city and its surrounding regions, the expansion of urban areas and overpopulation caused increased traffic and consequently other social problems, including traffic congestion and air pollution. To minimize such social costs, various policies to facilitate public transportation and taxis have been actively promoted [4]. Despite the efforts, the current services are short in providing an equal level of mobility to all residents, requiring enhancement. In particular, in Seoul Metropolitan City, despite its excellent public transportation infrastructure, there is a constant inconvenience of users with a need for enhancing problems caused by congestion during rush hours and a shortage of ridesharing (including taxi) supply late at night on the outskirts of the city and in hilly areas [5].

Against this backdrop and with the recent introduction of the mobility as a service(MaaS) concept, efforts have been made to improve individuals' mobility and accessibility through the integrated use of shared and cutting-edge modes of transportation [6–8]. MaaS consists of various means of transportation, including public transportation and a mobile platform-based ridesharing service. However, the ridesharing service is usually owned by the private sector, oriented toward profitability. Therefore, the provision of such a service tends to be concentrated in areas with high demand. Even if the ridesharing service were to expand, those users living in areas with low demand, especially suburban areas, are expected to face inconvenient services, such as longer waiting times due to the lower availability of vehicles. Especially during late-night hours when public transportation no longer operates, there are only limited options available, thus leading to further exacerbation of the imbalance between supply and demand. Therefore, a solution for this problem is needed. In a previous study [9] on experiences with taxi service refusal, respondents answered that they had difficulties getting taxi services or were refused during late-night hours (48.7%) when their destinations were either remote areas or outside of the city boundary (32.8%). From the perspective of users, this is recognized as a serious problem. The purpose of this study was to develop a dynamic pricing scheme to attract ridesharing (including taxi) drivers to mobility-disadvantaged regions during late-night hours when public transportation services are either reduced or come to a halt.

In the meantime, with the recent Fourth Industrial Revolution, the overall social paradigm has changed. A hyperintelligent society based on cutting-edge technologies, such as artificial intelligence, is being implemented. In the transportation sector, studies related to the introduction of big data- and artificial intelligence-based technology are growing [10]. Indeed, overseas ridesharing platforms, namely, Uber and Lyft, optimize supply and demand, as well as profitability, through AI-based dynamic pricing to retain the number of vehicles needed to respond to high demand in certain regions. Among the various subcategories of artificial intelligence, reinforcement learning has advantages in exploring unknown areas and identifying optimal outcomes through repeated exploration, unlike supervised learning, which is based on already available data. In contrast to other countries, there are not enough data related to ridesharing services in Korea and, accordingly, reinforcement learning would be more suitable. Consequently, this study conducted a reinforcement learning method using Seoul's taxi data to determine regionally appropriate levels of dynamic ridesharing fare rates with the purpose of improving the supply of ridesharing services in the mobility-disadvantaged regions in Seoul.

The paper is organized as follows: Section 2 reviews studies on the spatial fairness issues of transportation services and dynamic pricing applications for ridesharing services. Section 3 discusses the analysis framework and reinforcement learning method applied in this study. Subsequently, the results of the analysis are presented and discussed in Section 4, while the conclusions are drawn in Section 5.

2. Literature Review

2.1. Studies Analyzing Spatially Marginalized Areas in Terms of Transportation Services

In their study, Lee et al. [11] identified the mobility of different marginalized groups using smart card data and evaluated the mobility of groups highly reliant on public transportation. Building upon this, the study also categorized regions into different types according to their need for mobility improvement and concluded that mainly the outskirt areas of a city require improvements. Ha et al. [12] located areas marginalized from public transportation services in Seoul, using real travel characteristics gathered using Google and T map navigation application programming interfaces (APIs), and they analyzed areas with high priority in service enhancement. Data, such as travel characteristics and socioeconomic indices, were analyzed, identifying Gangbuk-gu, Seongbuk-gu, Seodaemun-gu, Jungnang-gu, and the southeast zone as areas requiring improvements in public transportation. Han [13] deduced spatially marginalized areas by assessing user mobility and the service level of providers. Moreover, the study reviewed the potential equity issue caused among different groups and suggested ways to improve it. According to the study, overall public transportation showed a satisfactory level from the users' perspective; however, there was a large deviation in terms of the supply level by region. Notably, on the outskirts of the city, including Nowon-gu and Gwanak-gu, the gap between supply and demand appeared to be wide and, thus, most urgently requiring an improvement in supply.

From an accessibility perspective, Lee et al. [14] developed various indices for the connectivity, directness, and diversity of public transportation using transportation card data and assessed each transportation zone. This study confirmed that, with a higher connectivity of public transportation, it had more routes and better directness. On the contrary, in zones located on the outskirts of the city, marginalization from public transportation was vivid. Kim et al. [15] and Yoon et al. [16] studied regional marginalization and inequity by considering not just spatial accessibility, but also social classes. Kim et al. overlaid and compared socioeconomic characteristics and city zones using location data of public transportation in Daegu. According to their analysis, the low accessibility and environmental inequity of socially disadvantaged people (the aged, recipients of national basic livelihood benefits, etc.) in suburban areas were confirmed. Yoon et al. calculated the inequity index of socially marginalized people on the basis of a Gini-style index and the methods of accessibility measure developed by Curie. As for public transportation accessibility, the regional gap was bigger for the subway than the bus. When this was overlaid on top of the data for the socially disadvantaged group, inequity was confirmed to be greater for the subway than the bus.

There were studies assessing equity depending on regional differences in infrastructure. Kim et al. [17] analyzed disadvantaged regions by overlaying service areas of public transportation and confirmed that suburb areas were mainly in a disadvantageous position. Furthermore, when considering socioeconomic characteristics, the study concluded that there was a gap in public transportation infrastructure among regions. Lee et al. [18] used Seoul Metropolitan Household Travel Survey data to measure regional equity among different income brackets depending on the levels of transportation infrastructure. The study showed that lower spatial equity led to longer total traveling time. Bin et al. [19] carried out a spatial cluster analysis at the administrative unit (Eup, Myeon, and Dong) level with transportation infrastructure indicators and travel behavior to assess equity in Gyeonggi-do province (excluding Seoul Metropolitan City and Incheon City). The results clearly showed gaps between areas closer to Seoul and those on the outskirts of the city. In particular, equity at the infrastructure level in northern Gyeonggi-do province was low.

2.2. Dynamic Pricing Studies on Ridesharing Service

Before reviewing previous literature, dynamic pricing can be defined as a strategy in which prices change flexibly for the same product or service depending on the market situation [20–22]. This strategy is mainly employed with respect to e-commerce, flight tickets, and hotel booking and demand management. This results in the optimization of selling products and services in an environment where the price can be easily adjusted. With regard to dynamic pricing for ridesharing services, there were studies on profitability improvement and the determination of an appropriate price through a pricing strategy [23–30], studies analyzing the elements of a pricing strategy affecting customers or drivers [31–33], and a study based on reinforcement learning [34].

First, Banerjee et al. [23] validated the performance of dynamic pricing by suggesting a ridesharing model considering two aspects: the stochastic dynamics of the market and the strategic decisions of the drivers, passengers, and platform. According to the analysis, depending on supply and demand conditions, flexible pricing resulted in increased total utility. Zeng et al. [24] researched the dynamic pricing strategy in accordance with potential users considering the destination of taxies. Markov Decision Process (MDP) was established by considering the cost of pick-ups at the destination. The total utility was enhanced compared to the fixed cost case. Hall et al. [25] studied the economic utility of surge pricing by analyzing Uber data.

When prices rise in line with the surge price algorithm, the delta affected the improving profitability of drivers, supply, and efficiency. Moreover, if surge pricing was not employed during peak hours, passenger waiting time was extended as drivers did not pick up passengers and passenger utility dropped.

In an analysis of the effects of dynamic pricing on drivers through Uber cases, Chen et al. identified that the surge price has a negative impact on passengers and a positive impact on drivers. Furthermore, the study found unfairness along the regional border of surge pricing. Chen et al. [32] studied changes in the number of Uber drivers depending on changes in surge price. The analysis confirmed that there was a higher rate of vehicle operation, as well as changes in operating hours, during the time period in which higher profit was expected from the drivers' perspective. Kooti et al. [33] analyzed the impact of dynamic pricing and income on the behavior of drivers, and they found that drivers operating vehicles during peak hours earned a higher income compared to nonpeak-hour drivers.

Wu et al. [34] simulated the application of dynamic pricing to ridesharing services. They compared four pricing methodologies, namely, (1) statistic pricing, (2) proportional pricing, (3) batch updates, and (4) reinforcement learning with the goal of profit maximization in a single Origin-Destination(OD) scenario. The simulation found that pricing based on reinforcement learning increased the total profit and individual profit of drivers the most.

2.3. Summary of the Review

Previous studies on mobility-disadvantaged regions mainly focused on evaluating areas marginalized from public transportation from a user's perspective utilizing big data, such as smart card data, household travel surveys, and Geographic Information System Data Base (GIS DB: location data for bus and subway, thematic transportation map, etc.). These studies identified that public transportation services in these regions were not adequately provided. On the other hand, they paid little attention to the late-night mobility inconvenience in regions after public transportation services had ended.

As for dynamic pricing in ridesharing services, studies were carried out mainly using Uber cases. As Uber discloses operation data, several studies related to dynamic pricing were conducted. In particular, the surge-based pricing analysis of Uber showed that it helped increase driver supply and operational profit. On the contrary, if a single-fare system was applied without flexibility, overall utility decreased due to a lower rate of matching and a longer waiting time when suppliers decided not to take passengers. However, there were no studies measuring regional differences in dynamic pricing; thus, it is necessary to determine ways of addressing drivers' refusal to take passengers when taking regional differences into account.

Accordingly, a study evaluating the dynamic pricing solution is required to improve the quality of mobility services in disadvantaged regions in Seoul.

3. Analysis Methodology

To simulate dynamic pricing using reinforcement learning, this study first conducted a centrality (in-bound and out-bound) analysis with a district(gu)-level OD matrix to develop a regional indicator of degree centrality to be used in reinforcement learning analysis. Then, the indicators were applied to a reinforcement learning simulation.

3.1. Analysis Methodology

3.1.1. Degree Centrality Analysis Methodology

Degree centrality is an indicator showing how many nodes are connected to a certain node. In social network theory, degree centrality is defined as the number of nodes linked directly to any given node. This study measured the out-degree and in-degree of each region (node) from a centrality perspective depending on the level of node connectivity, using them as indicators to be applied in reinforcement learning. Centrality can be calculated as follows:

$$C'_D(N_i) = \frac{\sum_{j=1}^{g} x_{ij}}{g-1}, \ i \neq j, \tag{1}$$

where C'_D is the standardized degree centrality of node i, $\sum_{j=1}^{g} x_{ij}$ is the degree centrality of node j, and g is the number of nodes. If there is no direction in the network, the above equation simply shows the degree of nodes. On the other hand, if directions are present, the equation distinguishes out-degree centrality (C'_{outD}) and in-degree centrality (C'_{inD}). Here, out-degree centrality is defined as the level of connections going out from a certain node to other nodes, and in-degree centrality is defined as the level of connections coming in from other nodes to a certain node. In this study, in-degree centrality referred to the number of vehicles coming into a certain zone (Gu), thus indicating a region as a frequently selected destination by passengers (usually residential areas during late-night periods). On the contrary, out-degree centrality referred to the number of vehicles traveling out of a certain zone, thus indicating a region as a preferred destination by drivers (central and subcentral regions). Therefore, this study identified indicators by region considering both out-degree and in-degree centrality.

3.1.2. Reinforcement Learning Methodology

Reinforcement learning is a learning method through which an agent chooses an action to take in an environment to maximize reward. This affects not only the immediate reward due to the action of the agent, but also the long-term reward. The main characteristics of reinforcement learning are a trial-and-error search and delayed reward (Figure 1). This study used the Q-learning algorithm for analysis.

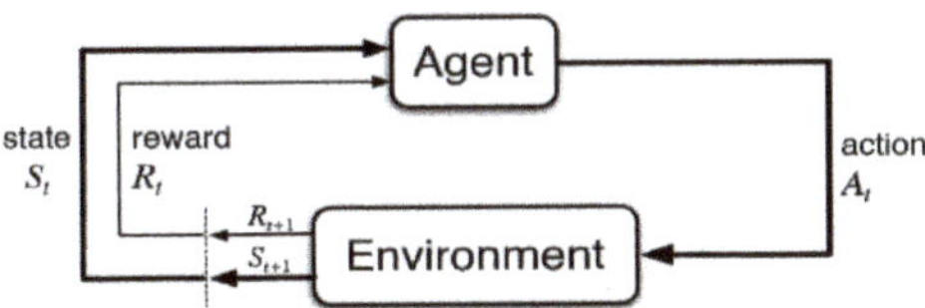

Figure 1. Conceptual diagram of reinforcement learning.

In the Q-learning algorithm [35,36], Q initially has an arbitrary fixed value. When the agent selects an action (a) in accordance with a learning step (t), an immediate reward (r) is observed before entering a new state (s), updating the Q value. The key characteristic of this algorithm is the value iteration update, using the weighted average of the old value and new information. The equation of Q-learning can be expressed as follows:

$$Q(s_t, a_t) \leftarrow (1-\alpha)Q(s_t, a_t) + \alpha(r_t + \gamma \max_{\alpha} Q(s_{t+1}, a), \tag{2}$$

where α is the learning rate which updates the current Q value with immediate reward and/or future expected value, usually between 0 and 1. When the value is 0, the current Q value is continuously used without any update according to learning. On the contrary, if the value is 1, the previous Q value is ignored and updated automatically. γ is a discount factor which is a variable explaining the difference between immediate and future rewards, also between 0 and 1. When the value is 0, the agent takes a myopic action as by making an update with immediate reward. When the value is 1, the agent leans toward a future reward, underestimating the current reward. In this study, on the basis of existing studies, we experimentally set the learning rate to 0.1 and discount factor to 0.9.

3.2. Environment Setting for Reinforcement Learning Simulation

3.2.1. Q-Table Configuration

To carry out the reinforcement learning simulation using the Q-learning algorithm, a Q-table needed to be defined through state and action. The Q-table is a matrix that compiles rewards for all states and actions, and it is updated after each learning step. Here, 600 OD cells were used in an OD matrix of 25 zones in Seoul excluding intrazonal travel. The state refers to the price that can potentially be generated in each OD cell depending on time slots, which can be expressed as 13 (price range) × 6 h (time round, Tr) for each OD cell. In total, there were 13 actions, including a standard price and surge prices from 0.6-fold to 3.0-fold the standard price with an interval of 0.2. The composition of the district OD matrix is shown in Figure 2, with a separate layer constructed for each cell's travel time and passage cost (P_{base}) to be used for analysis.

	Z1	Z2	Z3	Z4	Z5	Z6	Z7	Z8	Z9	Z10	Z11	Z12	Z13	Z14	Z15	Z16	Z17	Z18	Z19	Z20	Z21	Z22	Z23	Z24	Z25
Z1		0	1	2	3	4	5	6	7	8	9	10	11	12	13	14	15	16	17	18	19	20	21	22	23
Z2	24		25	26	27	28	29	30	31	32	33	34	35	36	37	38	39	40	41	42	43	44	45	46	47
Z3	48	49		50	51	52	53	54	55	56	57	58	59	60	61	62	63	64	65	66	67	68	69	70	71
Z4	72	73	74		75	76	77	78	79	80	81	82	83	84	85	86	87	88	89	90	91	92	93	94	95
Z5	96	97	98	99		100	101	102	103	104	105	106	107	108	109	110	111	112	113	114	115	116	117	118	119
Z6	120	121	122	123	124		125	126	127	128	129	130	131	132	133	134	135	136	137	138	139	140	141	142	143
Z7	144	145	146	147	148	149		150	151	152	153	154	155	156	157	158	159	160	161	162	163	164	165	166	167
Z8	168	169	170	171	172	173	174		175	176	177	178	179	180	181	182	183	184	185	186	187	188	189	190	191
Z9	192	193	194	195	196	197	198	199		200	201	202	203	204	205	206	207	208	209	210	211	212	213	214	215
Z10	216	217	218	219	220	221	222	223	224		225	226	227	228	229	230	231	232	233	234	235	236	237	238	239
Z11	240	241	242	243	244	245	246	247	248	249		250	251	252	253	254	255	256	257	258	259	260	261	262	263
Z12	264	265	266	267	268	269	270	271	272	273	274		275	276	277	278	279	280	281	282	283	284	285	286	287
Z13	288	289	290	291	292	293	294	295	296	297	298	299		300	301	302	303	304	305	306	307	308	309	310	311
Z14	312	313	314	315	316	317	318	319	320	321	322	323	324		325	326	327	328	329	330	331	332	333	334	335
Z15	336	337	338	339	340	341	342	343	344	345	346	347	348	349		350	351	352	353	354	355	356	357	358	359
Z16	360	361	362	363	364	365	366	367	368	369	370	371	372	373	374		375	376	377	378	379	380	381	382	383
Z17	384	385	386	387	388	389	390	391	392	393	394	395	396	397	398	399		400	401	402	403	404	405	406	407
Z18	408	409	410	411	412	413	414	415	416	417	418	419	420	421	422	423	424		425	426	427	428	429	430	431
Z19	432	433	434	435	436	437	438	439	440	441	442	443	444	445	446	447	448	449		450	451	452	453	454	455
Z20	456	457	458	459	460	461	462	463	464	465	466	467	468	469	470	471	472	473	474		475	476	477	478	479
Z21	480	481	482	483	484	485	486	487	488	489	490	491	492	493	494	495	496	497	498	499		500	501	502	503
Z22	504	505	506	507	508	509	510	511	512	513	514	515	516	517	518	519	520	521	522	523	524		525	526	527
Z23	528	529	530	531	532	533	534	535	536	537	538	539	540	541	542	543	544	545	546	547	548	549		550	551
Z24	552	553	554	555	556	557	558	559	560	561	562	563	564	565	566	567	568	569	570	571	572	573	574		575
Z25	576	577	578	579	580	581	582	583	584	585	586	587	588	589	590	591	592	593	594	595	596	597	598	599	

Zone	Name
Z1	Jongno-gu
Z2	Jung-gu
Z3	Yongsan-gu
Z4	Seongdong-gu
Z5	Gwangjin-gu
Z6	Dongdaemun-gu
Z7	Jungnang-gu
Z8	Seongbuk-gu
Z9	Gangbuk-gu
Z10	Dobong-gu
Z11	Nowon-gu
Z12	Eunpyeong-gu
Z13	Seodaemun-gu
Z14	Mapo-gu
Z15	Yangcheon-gu
Z16	Gangseo-gu
Z17	Guro-gu
Z18	Geumcheon-gu
Z19	Yeongdeungpo-gu
Z20	Dongjak-gu
Z21	Gwanak-gu
Z22	Seocho-gu
Z23	Gangnam-gu
Z24	Songpa-gu
Z25	Gangdong-gu

Figure 2. OD matrix, where each OD cell is the location data for each OD movement in the matrix.

3.2.2. Action Model

The simulation included passengers and drivers for each cell in the OD matrix and determined whether or not a match was feasible on the basis of the choice probability of passengers and drivers. Applying the method used by Wu et al. [34], passengers selected a price according to the surge according to a normal distribution. On the other hand, drivers applied choice probability according to a normal distribution with the weight of the centrality analysis outcome in accordance with their destination. Centrality analysis was applied to taxi operation data on the basis of 150 M links provided for Seoul Metropolitan City. The analysis was conducted using data from April to September of 2017 which were processed to generate OD data for each district (gu).

For the choice probability model of drivers and passengers, a surge of 1 (base fare of taxi) was applied by consulting the result of National Bureau of Economic Research(NBER, 2016) study [37]. In addition, the probability followed a cumulative normal distribution using the following equation:

$$\Phi(x) = \int_{-\infty}^{x} N(0,1)(x)dx. \tag{3}$$

The choice probability models of passengers (Equation (4)) and drivers (Equation (5)) according to the surge are expressed below.

$$P(accept|P_{offer}) = \Phi\left(\frac{P_{offer} - P_{base}}{\sigma}\right) \times W_1, \tag{4}$$

$$P(accept|P_{offer}) = \Phi\left(\frac{P_{base} - P_{offer}}{\sigma}\right) \times W_2 + \left(\frac{T_rC'_{outD}(N_j)}{T_rC'_{inD}(N_j)}\right) \times W_3, \tag{5}$$

where W_1 is the weight modifying the choice probability of passengers when surge = 1, W_2 is the weight modifying the choice probability of drivers when surge = 1, $T_rC'_{outD}(N_j)$ is the out-degree centrality value of a specific time round (T_r), $T_rC'_{inD}(N_j)$ is the in-degree centrality value of a specific time round (T_r), and W_3 is the weight of the equity index allowing differentiation by region.

3.2.3. Reward Function

The criteria for reward were different depending on matching. Once a driver was matched with a passenger, the reward was the base fare (P_{fare}) for the travel between origin and destination multiplied by the surcharge coefficient (S_a). If a driver was not matched with a passenger, a negative reward was applied in line with the waiting time value (W_t). The waiting time was set to 8 min considering the fact that the average waiting time during late-night periods when a surcharge is applied is 8.1 min according to a study in Seoul Metropolitan City [38]. The time value (V_t) was calculated on the basis of the taxi fare for 1 min in the OD matrix. While it increased profitability through a surcharge, it was also designed to provide a negative reward when unmatched by multiplying the time value for 8 min by the surcharge. The equation for the reward can be expressed as shown in Equation (6). Learning was conducted separately for when a negative value was applied to waiting time and for when it was not. Through a comparison, the study identified the appropriate reward function.

$$r = \begin{cases} matched & P_{fare} \times S_a \\ unmatched & \begin{cases} -W_t \times V_t \times S_a \rightarrow alt1) \\ 0 \qquad\qquad\quad \rightarrow alt2) \end{cases} \end{cases} \tag{6}$$

3.2.4. Q-Learning Algorithm

A total of 20,000 Q-learning reiterations were conducted by taking the index value for each zone into account in the late-night period (Figure 3). In one episode, the goal was to identify the optimal surge for each time slot during 6 h of operation for each OD matrix. The Q-table included states (current state and price) for each time slot and action value by applying the concept of time. The travel time and price required to calculate the reward value (operating profit) for each state referred to a separate OD table. The weight values applied to the preference of drivers were the centrality indices for each time slot and zone. The reward function was created separately depending on the application of the time value.

> *Algorithm parameters: $\alpha = 0.1, \gamma = 0.9$, epslion decay $= 0.995$*
> *Loop for each episode:*
>   *Initialize OD state, as od cell $\in$ od matrix*
>   *Loop for each OD state:*
>   *Initialize S, Cumulative time*
>   *Repeat:*
>   *Choose an action(A) from S using policy derived from Q (e.g., Decaying ε-greedy)*
>   *Take action (A), obtain reward R from reward function, and next state S'*
> *Update $Q(S, A) \leftarrow (1 - \alpha)Q(S, A) + \alpha(R + \gamma \max_{\alpha} Q(S', A)$*
> *Set $S \leftarrow S'$*
>   *Cumulative Time = Cumulative Time + time*
>   *Until cumulative time less than 6 hour*

Figure 3. Q-learning algorithm pseudo code.

4. Analysis Results

4.1. The Analysis Result of Areas Expected to Be Marginalized in Ridesharing Services

In order to identify the indices to be applied to reinforcement learning, the taxi operation data in Seoul were categorized into different zones according to 25 districts (gu), and the OD matrix was formed on the basis of the volume of pick-ups and drop-offs using district codes. Consequently, the study analyzed centrality and tried to identify a standardized index for each zone to be applied to reinforcement learning. The origin (the number of pick-ups) was compiled by matching administrative districts through the GIS spatial join function, while the destination (the number of drop-offs) was compiled by using the district(gu) code information for each vehicle departure. Any data including areas surrounding Seoul were excluded.

Using the OD matrix, degree centrality analysis for each zone was conducted, and the average in-degree and out-degree centrality was calculated for weekdays and weekends from 10:00 p.m. to 4:00 a.m. (Figures 4 and 5). According to the analysis, in-degree centrality displayed similar patterns for both weekends and weekdays, whereas the regional deviation was smaller than seen for out-degree centrality. Out-degree centrality was high in the Central Business Districts(CBDs) on weekdays and similarly high on weekends.

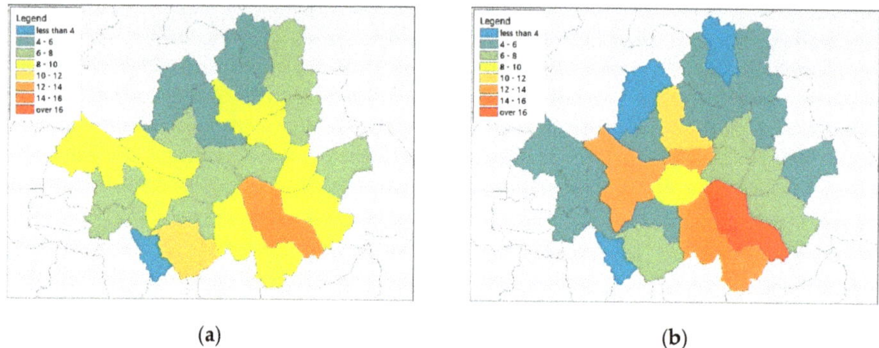

(a)   (b)

Figure 4. Results of centrality analysis between 10:00 p.m. and 4:00 a.m. on weekdays: (a) average in-degree centrality; (b) out-degree centrality.

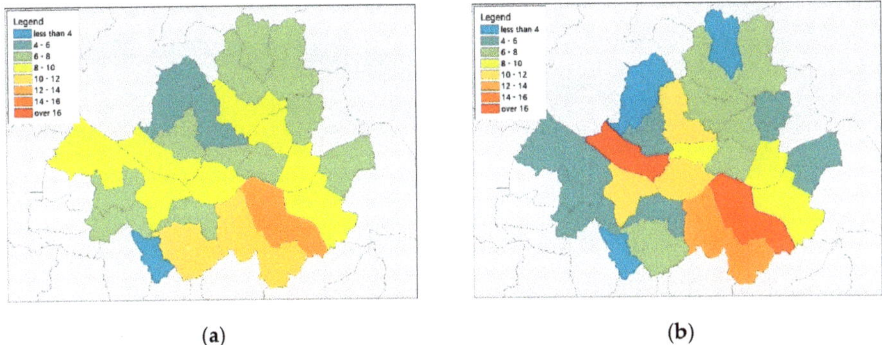

(a)   (b)

Figure 5. Results of centrality analysis between 10:00 p.m. and 4:00 a.m. on weekends: (**a**) average in-degree centrality; (**b**) out-degree centrality.

Areas located on the outskirts of the city had a lower degree centrality. In particular, outskirt areas showed lower out-degree centrality on weekdays. Residential areas on the outskirts of Seoul showed higher in-degree centrality. On the other hand, central and subcentral regions of the city showed higher out-degree centrality (Figure 6). The patterns were similar to those found in previous studies showing the concentration of taxi pick-ups during late-night hours near CBDs [39]. There were large differences in out-degree centrality by zone before and after midnight when most public transportation services were suspended, with the difference growing as the night went on.

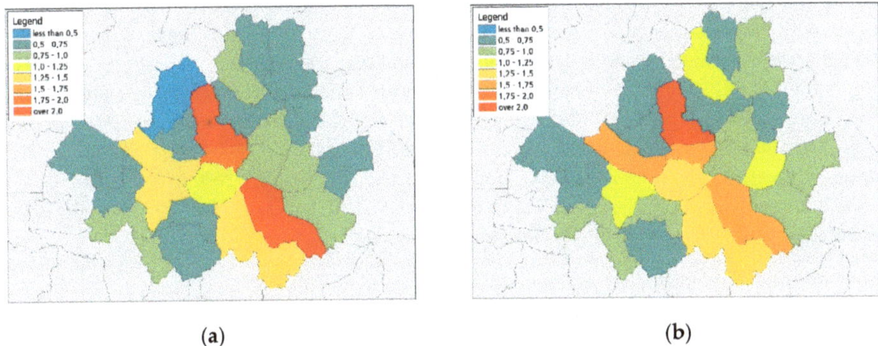

(a)   (b)

Figure 6. Results of out-degree centrality versus in-degree centrality analysis between 10:00 p.m. and 4:00 p.m.: (**a**) weekday; (**b**) weekend.

4.2. Results of Reinforcement Learning Simulation

The algorithm presented in Figure 3 was used in the learning of 600 OD cells. When the reward function was defined by a simple matching rate, there was a limitation in comparison among regions as the fare and travel time for each OD were different. Therefore, the simulation implemented a reward value considering travel time and cost. Moreover, analysis was conducted depending on the application of a negative reward value in the unmatched condition. The matching rate for each learning step when a waiting time with a negative reward was applied (alt1) or not (alt2) is depicted in Figure 7.

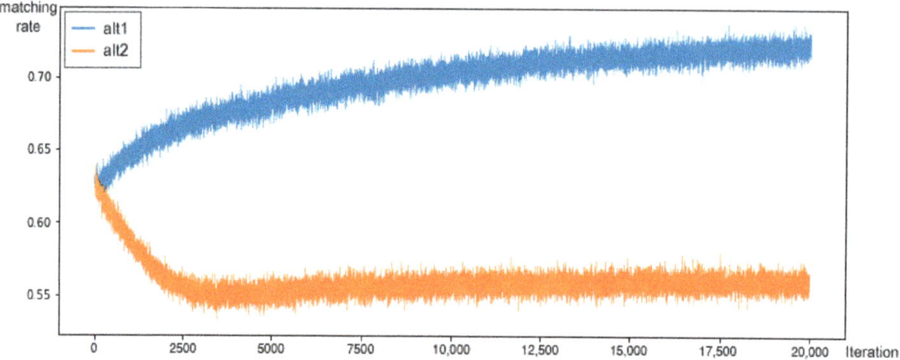

Figure 7. Matching rate comparison depending on time value.

When the negative reward was considered, the matching rate increased before converging as learning was reiterated. When it was not considered, the matching rate decreased before converging. According to this result, it can be concluded that the application of a negative reward was necessary when the reward function used the travel cost but not the matching rate. However, the current negative reward for the waiting time used a fixed value from a previous study; thus, for real-world applications, the expected waiting time according to a driver's choice probability should be calculated for each OD.

Using the deduced average surge, when a fare was calculated for each OD, the distribution could be expressed as shown in Figure 8, where the *x*-axis is the travel distance between origin and destination, and the *y*-axis is the price. Then, price levels were compared in terms of the base fare, base fare with late-night surcharge (additional 20% to base fare), late-night surcharge + ride-hail fee (KRW 3000, the highest fee charged by existing platform companies), and surge price (the outcome of learning for each OD). For travel distances shorter than 10 km, the differences among fares were not severe. However, the gap grew wider when the distance increased (see Figure 8a). This is because the late-night surcharge + ride-hail fee (yellow dotted line) was a flat fare regardless of traveling distance, while the surge fare (blue dotted line) was determined using a certain ratio, increasing the absolute price with traveling distance. Table 1 displays the three average prices for different distances, showing the price gap widening as the distance increased.

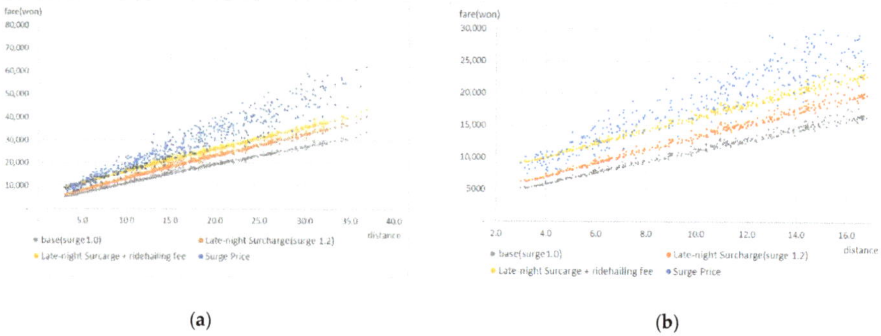

(a)   (b)

Figure 8. Price distribution for different OD when a surge was applied (using the average surge for different OD): (**a**) total range; (**b**) short distance.

Table 1. Traffic volume and average price (KRW) for difference distances.

Distance	Traffic Volume	Number of OD Zones	Surge Price	Late-Night Surcharge [1]	Late-Night Surcharge + Ride-Hail Fee [1]
Shorter than 5 km	28.5%	46	9494.6	7017.3	10,017.3
5–10 km	38.5%	116	14,845.0	10,450.8	13,450.8
10–15 km	18.4%	146	22,264.0	15,846.3	18,846.3
15–20 km	8.5%	108	29,625.4	20,863.1	23,863.1
20–25 km	4.2%	86	36,244.8	25,203.1	28,203.1
Longer than 25 km	1.8%	99	45,761.9	31,731.4	34,731.4
Total	100.0%	600	27,078.1	19,006.2	22,006.2

[1] The late-night surcharge is surge 1.2 of the base amount and an additional Hail Fee of 3000 KRW (based on the existing platform price).

4.3. Changes in Centrality after Surge-Driven Supply Increase

Centrality analysis was carried out after recalculating the traffic volume in accordance with the surge previously identified for each OD matrix and time slot. The traffic volume was recalculated as follows:

$$Nvol_{i,j} = \left(Ovol_{i,j} \times S_{i,j}\right) \times \frac{\sum_{i,j} Ovol_{i,j}}{\sum_{i,j} Ovol_{i,j} \times S_{i,j}}, \tag{7}$$

where i, j are the origin and destination, $Nvol_{i,j}$ is the recalculated traffic volume, $Ovol_{i,j}$ is the previous traffic volume, $S_{i,j}$ is the optimal surge for travel (optimal surge for different OD as identified from reinforcement learning), and $\sum_{i,j} Ovol_{i,j}$ is the sum of traffic volume. In centrality analysis, the indicator evaluating the volume of vehicles traveling to a certain zone was the in-degree centrality. When in-degree centrality increases, it can be said that supply is increasing toward that region.

According to the level of improved spatial equity by region, in-degree centrality decreased in the central and subcentral regions of the city, such as Gangnam-gu and Jongno-gu, while the in-degree centrality increased in residential areas on the outskirts of the city, such as Songpa-gu, Gangseo-gu, Dongjak-gu, and Nowon-gu (Figure 9). When this was overlaid on top of the previous hotspot analysis, it was found that the in-degree centrality was drastically improved in districts (gu) located on the outskirts of Seoul Metropolitan City where the number of drop-offs or vacant vehicles was greater compared to the number of pick-ups (Figure 10b). As a result, it can be expected that vehicle supply would be enhanced by as much as 7.5% when applying a higher surge to the region predicted to have a lower choice probability from the perspective of drivers. At the same time, equity in service supply would be improved by reducing the waiting time of passengers in marginalized regions with low taxi demand.

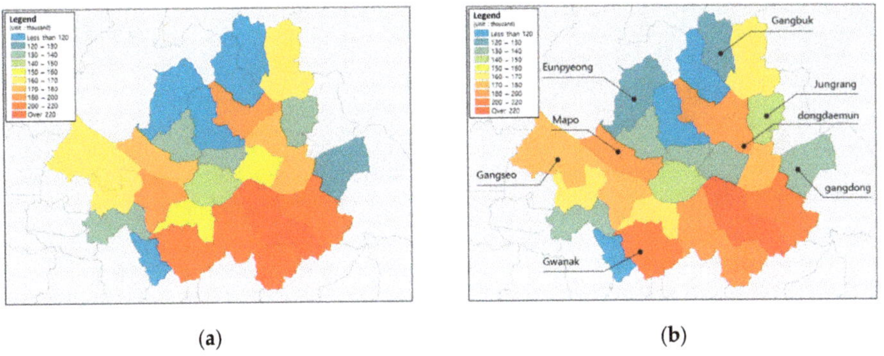

(**a**)   (**b**)

Figure 9. Change in in-degree centrality: (**a**) before; (**b**) after.

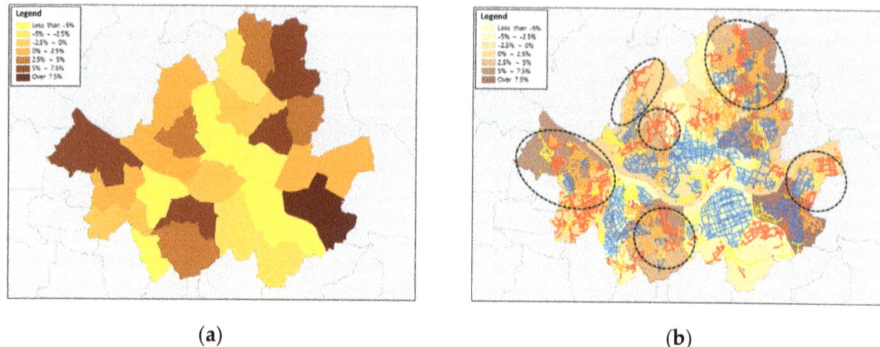

(a)   (b)

Figure 10. (**a**) Rate of change in in-degree centrality; (**b**) comparison between in-degree centrality and hotspot analysis.

5. Conclusions

As ridesharing (including taxi) services are often run by private companies, profitability is the top priority in operation. This leads to an increase in drivers' refusal to take passengers to areas with low demand where they have difficulties finding subsequent passengers, causing problems such as extended waiting time when hailing a vehicle for passengers bound for these regions. This problem differs depending on time and region. In late-night hours and in suburban regions, the imbalance between supply and demand is especially widened, worsening the problem. In order to address this problem, solutions were proposed in this study through a dynamic pricing strategy using reinforcement learning algorithms.

This study used Seoul city's taxi data to find appropriate fare surge rates for ridesharing services between 10:00 p.m. and 4:00 a.m. In reinforcement learning, the outcome of centrality analysis was applied as the weight affecting drivers' destination choice probability. Moreover, the reward function used during learning was adjusted according to whether or not a passenger waiting time value was applied. Profit was used as the reward value. By applying a negative reward for the passenger's waiting time, a more appropriate surge fare level could be identified. Across the region, the average surge level amounted to 1.6. Regions located on the outskirts of the city in predominantly residential regions such as Gangdong-gu, Dongjak-gu, Eunpyeong-gu, and Gangseo-gu showed a higher surge. On the contrary, central areas, such as Gangnam-gu, Jongno-gu, and Jung-gu, had a lower surge. The findings showed that the supply of ridesharing services in low-demand regions could be increased by as much as 7.5% using surge fares, thereby reducing regional equity problems related to ridesharing services in Seoul to a great extent.

This study conducted a reinforcement learning-based dynamic pricing simulation to respond to the regional equity problem of ridesharing (including taxi) services in Seoul. A novel approach was presented using dynamic pricing as a way to mitigate the spatial equity problem by affecting ridesharing supply, unlike most previous dynamic pricing studies which simply targeted higher profitability. Notably, it was shown that a surge rate change in fares could reduce the indirect refusal of drivers to take passengers to unpreferred areas. With additional real-time ridesharing user data, the Deep Q-Network(DQN) technique can be adopted to conduct a smaller-scale spatial analysis of ridesharing services. Furthermore, with more knowledge on fare sensitivity by user group, the dynamic pricing approach proposed in this study can significantly contribute to resolving the spatial equity problem in mobility services in the future.

Author Contributions: Conceptualization, K.Y.H.; data curation, J.S.; investigation, Y.J.C. and M.H.K.; methodology, J.S.; project administration, M.H.K.; software, J.S.; writing—original draft, J.S.; writing—review and editing, K.Y.H. All authors have read and agreed to the published version of the manuscript.

Electronics **2020**, *9*, 1818

Funding: This research was supported by a grant (20TLRP-B148970-03) from the Transportation and Logistics R&D Program funded by the Ministry of Land, Infrastructure, and Transportation of the Korean government.

Acknowledgments: This paper is a modification and expansion of Jae In Song's doctoral dissertation.

Conflicts of Interest: The authors declare no conflict of interest.

References

1. Park, J.; Kim, D.; Ko, Y.; Kim, Y.; Park, K. Urbanization and urban policies in Korea. *Korea Res. Inst. Hum. Settl.* **2010**, *2010*, 56.
2. CHO, M. Trend and prospect of urbanization in Korea: Reflections on Korean cities. *Econ. Soc.* **2003**, *60*, 10–39.
3. Lim, K. Characteristics of Urbanization and its Policies in Korea. *J. Korean Reg. Dev. Assoc.* **2004**, *16*, 139–156.
4. Hwang, S. Revitalizing the Transportation Market with the Public Transportation Promotion Act. *Korean Soc. Transp.* **2005**, *2*, 7–13.
5. Hong, S.; Yoon, H.; Yu, K.; Yun, J. A Study on Deriving of Vulnerable Sections in Public Transportation Services. *Seoul Inst.* **2019**, *2019*, BR-02.
6. Heikkilä, S. Mobility as a Service—A proposal for Action for the Public Administration, Case Helsinki. Master's Thesis, Aalto University, Aalto, Finland, 2014.
7. Cox, N.C.J. Estimating Demand for New Modes of Transportation Using a Context-Aware Stated Preference Survey. Ph.D. Thesis, Massachusetts Institute of Technology, Cambridge, MA, USA, 2015.
8. Finger, M.; Bert, N.; Kupfer, D. *Mobility-as-a-SERVICES: From the Helsinki Experience to a European Model?* European Transport; Regulation Observer: San Domenico di Fiesole, Italy, 2015.
9. Lee, J.; Kim, D. The role of smart mobility in resolving the imbalance in transportation supply and demand. In Proceedings of the 2018 Spring Academic Conference of the Korea Society of Management Information System, Seoul, Korea, 10 November 2018; pp. 139–158.
10. Kim, K.; Kim, T.; Cho, S.; Jang, H. *Deregulation Strategies for Promoting Innovation of Technology and Services in Transport Sector in the Era of 4th Industrial Revolution*; The Korea Transport Institute: Sejong City, Korea, 2019.
11. Lee, H.; Ha, J.; Lee, S. An Analysis on the Equity of Public Transit Service using Smart Card Data in Seoul, Korea—Focused on the Mobility of the Disadvantaged Population Groups. *J. Korean Reg. Sci. Assoc.* **2017**, *33*, 101–113.
12. Ha, J.; Lee, S. An Analysis of Vulnerable Areas for Public Transit Services using API Route Guide Information—Focused on the Mobility to Major Employment Centers in Seoul, Korea. *J. Korea Plan. Assoc.* **2016**, *51*, 163–181. [CrossRef]
13. Han, D. The Analysis of Public Transport Accessibility and Equity in Seoul. Ph.D. Thesis, Graduate School of Konkuk University, Seoul, Korea, 2016.
14. Lee, H.; Ko, S.; Kim, D. Development of Equity Indicators for Public Transportation Using Transportation Card Data. In Proceedings of the 2018 Spring Academic Conference of The Korean Institute of Intelligent Transport. Systems, Jeju-do, Korea, 19–21 April 2018; pp. 471–476.
15. Jun, B.; Kim, A. Environmental Equity Analysis of the Accessibility to Public Transportation Services in Daegu City. *J. Korean Assoc. Geogr. Inf. Stud.* **2012**, *15*, 76–86.
16. Yun, J.; Woo, M. Empirical Study on Spatial Justice through the Analysis of Transportation Accessibility of Seoul. *J. Korea Plan. Assoc.* **2015**, *50*, 69–85. [CrossRef]
17. Kim, J.; Kang, S.; Kwon, J. The Spatial Characteristics of Transit-Poors in Urban Areas. *J. Korean Assoc. Geogr. Inf. Stud.* **2008**, *11*, 1–12.
18. Lee, W.; Bin, M.; Moon, J.; Cho, C. Analysis and Policy Proposal on the Vulnerable Areas of Public Transportation Accessibility in Gyeonggi-do. *2012 Acad. Conf. J. Korean Soc. Transp.* **2012**, *67*, 47–50.
19. Bin, M.; Lee, W.; Moon, J.; Joh, C. Integrated Equity Analysis Based on Travel Behavior and Transportation Infrastructure: In Gyeonggi-Do Case. *J. Korean Soc. Transp.* **2013**, *31*, 47–57. [CrossRef]
20. Elmaghraby, W.; Keskinocak, P. Dynamic pricing in the presence of inventory considerations: Research overview, current practices, and future directions. *Manag. Sci.* **2003**, *49*, 1287–1309. [CrossRef]
21. Rouse, M. What is Dynamic Pricing?—Definition from WhatIs. com. WhatIs. com. 2015. Available online: https://whatis.techtarget.com/definition/dynamic-pricing (accessed on 1 November 2020).

22. den Boer, A.V. Dynamic pricing and learning: Historical origins, current research, and new directions. *Surv. Oper. Res. Manag. Sci.* **2015**, *20*, 1–18. [CrossRef]
23. Banerjee, S.; Johari, R.; Riquelme, C. Dynamic pricing in ridesharing platforms. *ACM SIGecom Exch.* **2016**, *15*, 65–70. [CrossRef]
24. Zeng, C.; Oren, N. *Dynamic Taxi Pricing*; IOS Press: Amsterdam, The Netherlands, 2014.
25. Hall, J.; Kendrick, C.; Nosko, C. *The Effects of Uber's Surge Pricing: A Case Study*; The University of Chicago Booth School of Business: Chicago, IL, USA, 2015.
26. Li, S.; Fei, F.; Ruihan, D.; Yu, S.; Dou, W. A dynamic pricing method for carpooling service based on coalitional game analysis. In *2016 IEEE 18th International Conference on High Performance Computing and Communications; IEEE 14th International Conference on Smart City; IEEE 2nd International Conference on Data Science and Systems (HPCC/SmartCity/DSS)*; IEEE: Piscataway Township, NJ, USA, 2016; pp. 78–85.
27. Waserhole, A.; Jost, V. Pricing in vehicle sharing systems: Optimization in queuing networks with product forms. *EURO J. Transp. Logist.* **2016**, *5*, 293–320. [CrossRef]
28. Qiu, H.; Li, R.; Zhao, J. Dynamic pricing in shared mobility on demand service. *arXiv* **2018**, arXiv:1802.03559. Available online: https://arxiv.org/pdf/1802.03559.pdf (accessed on 1 November 2020).
29. Korolko, N.; Woodard, D.; Yan, C.; Zhu, H. Dynamic Pricing and Matching in Ride-Hailing Platforms. 2018. Available online: https://api.semanticscholar.org/CorpusID:53059883 (accessed on 1 November 2020).
30. Avdeenko, T.; Khateev, O. Taxi Service Pricing Based on Online Machine Learning. In *International Conference on Data Mining and Big Data*; Springer: Singapore, 2019; pp. 289–299.
31. Chen, L.; Mislove, A.; Wilson, C. Peeking beneath the hood of uber. In Proceedings of the 2015 Internet Measurement Conference, Tokyo, Japan, 28–30 October 2015; pp. 495–508.
32. Chen, M.K.; Sheldon, M. Dynamic Pricing in a Labor Market: Surge Pricing and Flexible Work on the Uber Platform. UCLA Anderson. 2015. Available online: https://www.anderson.ucla.edu (accessed on 1 November 2020).
33. Kooti, F.; Grbovic, M.; Aiello, L.M.; Djuric, N.; Radosavljevic, V.; Lerman, K. Analyzing Uber's ride-sharing economy. In Proceedings of the 26th International Conference on World Wide Web Companion, Perth, Australia, 3–7 April 2017; pp. 574–582.
34. Wu, T.; Joseph, A.D.; Russell, S.J. *Automated Pricing Agents in the on-Demand Economy*; University of California at Berkeley: Berkeley, CA, USA, 2016; UCB/EECS-2016-57.
35. Sutton, R.S.; Barto, A.G. *Reinforcement Learning: An Introduction*; MIT Press: Cambridge, MA, USA, 2018.
36. Van Otterlo, M.; Wiering, M. Reinforcement learning and markov decision processes. In *Reinforcement Learning*; Springer: Berlin/Heidelberg, Germany, 2012; pp. 3–42.
37. Cohen, P.; Hahn, R.; Hall, J.; Levitt, S.; Metcalfe, R. Using Big Data to Estimate Consumer Surplus: The Case of Uber. Available online: https://www.nber.org/system/files/working_papers/w22627/w22627.pdf (accessed on 1 November 2020).
38. An, G.; Kang, T.; Kim, B. *Taxi Service Evaluation Service Report 2014*; Policy Report 186; Seoul Institute: Seocho-gu, Seoul, Korea, 2015.
39. An, G. *Seoul's Taxi Usage and Operation Status and Improvement Plan*; Policy Report 186; Seoul Institute: Seocho-gu, Seoul, Korea, 2015.

Publisher's Note: MDPI stays neutral with regard to jurisdictional claims in published maps and institutional affiliations.

MDPI

St. Alban-Anlage 66

4052 Basel

Switzerland

Tel. +41 61 683 77 34

Fax +41 61 302 89 18

www.mdpi.com

Electronics Editorial Office

E-mail: electronics@mdpi.com

www.mdpi.com/journal/electronics